# THE PIPELINE:

# A PICTURE OF HOMEBUILDING PRODUCTION

How RB Builders Learned to Apply
the Principles and Disciplines
Governing Homebuilding
Production.

## SECOND EDITION, 2016

Fletcher L. Groves, III

"The Pipeline: A Picture of Homebuilding Production, Second Edition," by Fletcher L. Groves, III. ISBN 978-1-62137-804-4.

First Edition published 2013, ISBN 978-1-62137-193.9.

THE PIPELINE: A PICTURE OF HOMEBUILDING
PRODUCTION, SECOND EDITION

## COMMENTS, BOOK RECOMMENDATIONS,
## AND BOOK REVIEWS

"First, congratulations on producing a powerful and engaging book that can have significant impact on the homebuilding industry. Second, the book holds the keys to systematic and continuous improvement within any industry. And, your shaping of the environment for the production homebuilder makes the discovery even more relevant. Third, you have captured what you set out to do.

"And, the conclusion of *The Pipeline* is beautiful in its simplicity: Discipline, Context, Perspective. Improving business performance boils down to getting the job done – viewing the issue, sustaining the effort, and getting the results – in three critical dimensions."

Hoyt G. Lowder
HGB & Associates, LLC
Tampa, FL
December 2012

(former Director at FMI Corporation)

---

"Although I am not from the homebuilding profession, I found *The Pipeline* to be instructional, informative, easy to read, and thoroughly entertaining. The dialog was believable, crisp, and to-the-point. I'm impressed with Fletcher's facility with words and the style he employed – like *The Goal* and *The Challenge,* two of my favorite books. Fletcher is a natural writer. *The Pipeline* was easier for me to read (than *The Goal* and *The Challenge*) because Fletcher used job titles rather that a person's given name. I did not have to stop and think what function the individual performed. Fletcher cleverly summarized the key concepts of each preceding chapter before introducing the new.

"I liked the erasable board and flipchart use of the intrepid, results-based consultant, as she documented key concepts and then referred back to them when necessary. I think using the CEO character to introduce, support, confirm, and explain – as necessary – company policy and direction, was a stroke of genius. We all benefit greatly when we can understand how the big boss thinks. That understanding provides immeasurable guidance to subordinate day-to-day choices.

"For me, a logistics professional and a practitioner in continuous process improvement, Chapter V: Production Processes was spot on. The discussions regarding systems, processes, variation, waste, Theory of Constraints, bottlenecks, buffers, pull v. push, balanced capacity v. unbalanced capacity, and etc. were important, striking, and remarkable. True genius, in my humble opinion!

"For those of you in the homebuilding profession, *The Pipeline* is a must read. For those of us in other professions, *The Pipeline* is an informative, thought-provoking, and fascinating work that helps us understand how a successful homebuilding company can improve to be best-in-class. There are many lessons that the reader can borrow from RB Builders and apply to his/her situation."

Joe Kinsey
President, Operating Results, Inc.
Orange Park, FL
January 2013

_____

"First and foremost, I salute my friend and colleague of so many years, Fletcher Groves, on his accomplishments as a husband, father and respected pillar of his community. He is a man of many achievements, but all laced with a common theme of professionalism and integrity.

"In writing _The Pipeline_, Fletcher has integrated what he has learned as a banker, homebuilder, process engineer and consultant in a variety of fields to create a practical (but interesting and amusing) handbook for homebuilders. Moreover, the approach and lessons drawn from a variety of relevant process and quality sources make this book a meaningful resource for any business that has a variety of processes, moving parts and dependencies crying for definition and management.

"_The Pipeline_ should also become a textbook for any building construction student and a resource for others engaged in the struggle for the definition of quality in business. Again, my congratulations on a brilliant work."

Kent Steen
Vice President – Marketing/Sales
Bitrage NOVIS Corporation
Warrior and Refugee from Service and Administrative Institute (SAI)
January, 2013

_____

"Other industries have been executing on improving business productivity, predictability, and profitability for years, with little of that learning making it to the homebuilding industry. When I was introduced to Fletcher Groves over a decade ago, he was one of the few who could introduce us to this thinking, and what he shared with us has become part of our culture, part of who we are.

"With _The Pipeline_, Fletcher has made it even easier to understand these ideas, taking a great body of knowledge and transforming it into a story directly applicable to homebuilding. You quickly get the perspective you need to secure a strong and defendable future for your business.

"If you want do more than survive in this business, read this book."

Robert Bowman
President
Charter Homes & Neighborhoods

January 2013

Charter Homes & Neighborhoods earned the 2013 GOLD National Housing Quality Award

---

"We would recommend *The Pipeline: A Picture of Homebuilding Production* to anyone just entering the homebuilding industry, or industry veterans that have been building homes for decades. The story of RB Builders breaks down the important drivers of profitability and return on investment into simple-to-understand components that can be influenced everyday by team members.

"Over the past decade, incorporating the concepts that Fletcher depicted in the narrative of RB Builders, Jagoe Homes has not only been able to survive the recession, but to strengthen our position in the marketplace and increase our profitability and return on investment. *The Pipeline: A Picture of Homebuilding Production* will provide its readers with the same tools and concepts that Fletcher helped us with at Jagoe Homes to improve our business results."

Scott and Brad Jagoe
Jagoe Homes Inc.
Owensboro, KY

Jagoe Homes was named 2010 Builder of The Year by *Professional Builder*.

---

"All homebuilders focus on margin. It's what we think about. It's what we benchmark about with other builders: 'What did you sell that house for? How much did you pay for that lot? What were your construction costs per sq. ft.?' What we need is the ability to more-fully understand the relationship between margin *and* velocity, as the co-determinants of economic return.

"In *The Pipeline: A Picture of Homebuilding Production*, Fletcher clearly explains how business outcomes are maximized, not simply by controlling and extracting maximum value from its direct variable costs – by increasing margin – but, also by leveraging the indirect, non-variable costs that represent the internal resources associated with its overhead, exploiting the external resources associated with its suppliers and subcontractors, and limiting the amount of work-in-process.

"Together, it is those last three efforts, together with managing production as a system, that determine the utilization of production capacity.

"That is what increases velocity."

Ben Walters
Chief Operating Officer,
Tennessee
Dock Street Communities, Inc.

---

| CONTENTS | PAGE |
|---|---|

# FOREWORD:

When my friend and colleague Fletcher Groves told me he was writing a book explaining homebuilding production principles and disciplines, I was pleased and supportive.

Fletcher is a man of energy, enthusiasm, and profound experience who has taught me a ton about this fascinating and essential industry (it's been my good fortune to work with Fletcher and the great Jack Suarez on the Inland Homebuilding System).

Fletcher is unique in that he combines a deep knowledge of Lean (also known as the Toyota Production System), Theory of Constraints, Business Process Management, and Finance – a powerful combination. He thereby avoids the tiresome "theological" debates – Lean vs. TOC vs. Business Process Reengineering vs. Six Sigma and so on – that distract and confuse. As Ernest Hemingway once observed, in a different context: It is all true. The point is to integrate powerful ideas toward achieving prosperity for our business, team members, and community.

He is also a man of decency and integrity which comes through in *The Pipeline*'s sub-theme of what the ancients called Fortitude – the guts to confront brutal facts without ever losing faith in the ultimate outcome.

I was lucky enough to grow up professionally at Toyota Motor Manufacturing Canada[1], and to spend extended periods of time at leading Toyota facilities in North America and Japan. Our (very patient) Toyota senseis emphasized the concept of a system – an organized set of parts with a clearly defined goal. Absent a system, at best, we sub-optimize; at worst, we waste our time.

And so, Fletcher has done us a great service. He has produced an engaging and accessible overview of what W. Edwards Deming has called the "profound system of knowledge", as it applies to homebuilding production. RB Builders, our fictional production homebuilding company, faces the very real challenges, and learns a powerful way of thinking and managing.

Homebuilding has endured a terrible downturn. But it will come back, as it always has. If we can learn and apply the principles explained in *The Pipeline*, if we can learn to think of homebuilding as a system, as RB Builders does in the story, we can blunt future boom-bust cycles, and thereby reduce human misery and preserve hard-won prosperity.

We're early on in this essential journey; a respected colleague suggested that, apart from a handful of progressive companies, homebuilding is where auto manufacturing was a century ago.

*The Pipeline: A Picture of Homebuilding Production* is an invaluable guide.

– Pascal Dennis

---

[1] Pascal Dennis is a professional engineer and President of Lean Pathways Inc., an international consultancy. He is a four-time winner of the Shingo Prize for Excellence. For more please visit Pascal's page at www.amazon.com. or www.leansystems.org

**PREFACE:**

This is the story about a mythical homebuilding company – one that we will simply call RB Builders – and how its cross-functional production team learned the principles of homebuilding production, with the assistance of its intrepid, results-based consultant.

Since it does not offer a central plot, *The Pipeline: A Picture of Homebuilding Production* is not a novel. It is a narrative, a story told in the exchanges of dialog between characters without names, characters who are identified by their respective functions and job descriptions; characters without names does not imply an absence of personality.

RB Builders is a production homebuilding company, so its lessons and efforts in understanding production extend from that context. RB's context of production might be different from, say, a custom homebuilding company, but the underlying principles and disciplines of production are universal; they are what we term production physics, and they have their roots in the laws that govern all manufacturing production.

In many respects, RB Builders is the ideal client for a consulting firm. Its owners are enlightened; its management is competent; its teammates are capable; and most importantly, it has problems to solve. The owners have recently chosen a new path, based on (1) an assessment of current reality; (2) a focused, targeted approach to continuous improvement; and (3) a commitment to a team-based method of performance compensation focused on a single, specific business outcome.

One of RB Builders' most pressing concerns – and one of its most significant limitations to improving business performance – is the lack of a clear, comprehensive understanding of production management.

That is what *The Pipeline* is about.

*The Pipeline* began life in 2007, at a relatively early point in the Great Housing Recession, as a way of explaining homebuilding production principles and disciplines through a course of dialog. It was written for a particular homebuilding client of SAI Consulting. The client eventually opted for a more traditional narrative, liberating me to write this book a bit differently.

*The Pipeline* took me five years to complete. The lessons the book contains goes back considerably further.

Almost twenty years ago, when I approached Service and Administrative Institute (as SAI was then known) about the use of its consulting platform in the homebuilding vertical, SAI was a national TQM firm focused on the transportation and logistics space, with a smaller group of engagement clients outside of that vertical in the Jacksonville, Florida market. That group included an aggregates company, an REIT, and other businesses interested in quality management; it also included a major Canadian steel mill.

In those days, Service and Administrative Institute was a State of Florida Sterling Award judge, that also provided quality management training for national companies, including Motorola, Milliken, BF Goodrich, Monsanto, and CSX Transportation.

With its focus on Total Quality Management, SAI was a local and logical platform conducive to where I wanted to focus my efforts. I had just concluded ten years in residential development and construction, including a stint with Arthur Rutenberg Corporation, all of which followed ten years in commercial banking. In some ways, the choice was simple: When you live in Ponte Vedra Beach, Florida, you tend to look for opportunities to work where you live, and avoid having to live where you work.

My joining SAI also coincided with the release of *Moving Towards The Future: A Builder's Introduction to Total Quality Management* by Gary Lewis, published through the NAHB Research Center. It marked the founding of the National Housing Quality award shortly thereafter.

As I labored to develop my own consulting practice focused on the homebuilding vertical, SAI assigned me to engagements with many of its other clients, from which I learned immensely. Those clients were gradually replaced with homebuilding clients. All of that work led to a focus on processes and workflow, in its various versions of Business Process Improvement (BPI), Business Process Management (BPM), and Business Process Reengineering (BPR). This work also lead directly into Lean Production/TPS, Six Sigma, and – most importantly – the Theory of Constraints (TOC).

These and other disciplines formed the basis of what became the firm's new consulting model, after the sale of SAI's transportation and logistics consulting practice to Trimac Logistics in 2000.

As the practice solidified and grew within the homebuilding vertical, almost all of my work came to involve business process mapping, almost always at an early stage of the client engagement. SAI became the industry's foremost practitioner of this narrow and specialized discipline. But – because process mapping focuses on the design, documentation, and improvement of the workflow central to creating the value that customers are willing to buy – these engagements also tended to drive the effort into other areas of needed improvement.

Over the years, these process mapping engagements have been the catalyst that has expanded SAI's work into other areas, including enterprise-wide operational assessments and scorecards, performance measurement and compensation systems, business literacy, and – for a number of clients – the design and codification of the entire production system.

That, too, is what *The Pipeline* is about.

*The Pipeline* is about a production system with an enduring visual image, the elements of which are crafted to the specific requirements of the homebuilding industry, and the entirety of which is always managed as a system. *The Pipeline* is about using the process management and project management tools that work, without regard to the consulting religion from which they come. *The Pipeline* makes the inherent, inviolable, real, and measurable connection between operating performance and business outcomes.

Grateful acknowledgements for the contributions and roles others played in the writing of *The Pipeline: A Picture of Homebuilding Production.*

The long list of thinkers and writers who were formative to my knowledge and understanding of the methods and management of systems, processes, and project portfolios, formative to my knowledge and understanding of strategy, compensation, and managerial accounting, including:

Eli Goldratt;  Jim Womack and Dan Jones;  Michael Treacy and Fred Wiersema;  Phillip Crosby;  James Harrington;  Dan Hunt;  Dan Madison;  Arthur Tenner and Irving DeToro;  Alec Sharp and Patrick McDermott;  Geary Rummler and Alan Brache;  Christopher Meyer;  Jerry Harbour;  Bruce Silver;  Jim Champy and Michael Hammer;  Jason Jennings;  Bill Jensen;  Jim Collins;  John Case;  Robert Schaffer and Ron Ashkenas;  David Maister, Charles Green, and Robert Galford;  John Kotter;  Dave Ulrich, Jack Zenger, and Norm Smallwood;  Merom Klein and Rod Napier;  Peter Pande, Robert Neuman, and Roland Cavanagh;  Michael George;  Mike Rather and John Shook;  Jeff Cox;  Jim Cox;  Rob Newbold;  Larry Leach;  Sebastian Nokes, Ian Major, Alan Greenwood, Dominic Allen, and Mark Goodman;  Bill Dettmer;  Lisa Scheinkopf;  Gerald Kendall; Dee Jacob and Suzan Bergland;  Mark Graham Brown;  Bill McGuinness;  Jean Cunningham and Orest Flume;  Thomas Corbett;  Joel Siegel and Jae Shim;  and Ray Garrison and Eric Noreen.

Cort Dondero, then-CEO of Service and Administrative Institute, who gave me the opportunity, and the freedom, to develop my own consulting practice;  my former colleagues and friends at SAI:  Kent Steen, Joe Kinsey, Bob Pues, and, particularly, Steve Hollwarth, who is now mapping business processes at a much higher level.

Mike Hollister, friend, president of Hollister and Associates, Inc., and consulting lead on some of the more instructive engagements with homebuilding clients;  Mike was an occasional co-presenter with me at IBS and occasional co-author of *Reference Point*®, SAI's C-Level management survey conducted periodically among the building companies on *Professional Builder's Annual Survey of Housing Giants.*

Pascal Dennis, another friend, a colleague at Lean Pathways, co-consultant on various engagements, Shingo Award-winning author of *Lean Production Simplified*, *Andy & Me*, and *The Remedy*, Poet Laureate of Traveling Consultants (Bloomberg/Business Week), author, as well, of his latest book, *Reflections of a Business Nomad: Stories and Poems From the Road.*

Pascal provided expert insight into Lean/TPS principles and the more-broadly applied area of what is termed Factory Physics, all the while, graciously and patiently enduring my production heresies.

Scott Sedam, president of TrueNorth Development, fellow consultant, fellow writer, who has – through determination and perseverance – succeeded in the adaption and practical application of Lean tools in an industry disparate from whence they originated.

For certain, several of the early-joiners in SAI's business process improvement work, the leaders whose homebuilding companies went on to win either the National Housing Quality (NHQ) Award, or to earn Builder of the Year recognition from *Professional Builder*, notably: Rob Bowman at Charter Homes & Neighborhoods; Bill and Scott Jagoe at Jagoe Homes.

There are few people in the homebuilding industry from whom I have learned more than my former boss, Art Rutenberg, the always-accessible (he taught me that, too) chairman of Arthur Rutenberg Homes, Inc.

I hope *The Pipeline* repays some of that debt, because ARH is a picture of consistency, discipline, and elegant systems integration on everything related to how much a franchisee makes on each home – and on absolutely nothing related to how many homes that franchisee can produce with a planned, finite, and controlled amount of production capacity.

Art and I agree that we live our lives on opposite sides of the DuPont formula; he is all about margin, and I am all about velocity. The reality is, a homebuilding company needs both.

He tells me my job is easy; I ask him, if what I do is so easy, why isn't he any good at it.

I hope Art enjoys reading *The Pipeline*.


Most of all, I want to thank Jack Suarez – the *great* Jack Suarez, as Pascal acknowledges – good and longtime Tampa friend, third-generation builder, founder and chairman of Inland Homebuilding Group.

Jack listened, reasoned, balanced competing influences, challenged, and pushed; he rejected business-as-usual, he invested in change, and, in so doing, he imposed operating discipline and focus. In the end, he entrusted me with enterprise-level responsibility and the freedom to design solutions for a range of initiatives that coalesced into the Inland Homebuilding System.

From that perspective, the thinking behind the work at IHG went far beyond production principles and disciplines, to the point of defining the underlying Inland business model. It also influenced the SAI consulting model. In that sense, *The Pipeline* is about more than production principles and disciplines; it is about what it takes to achieve the business outcomes – the results – that justify the importance of having a system of homebuilding production.

Absent the influence of Jack Suarez, *The Pipeline: A Picture of Homebuilding Production* probably would not have been written.

– Fletcher Groves, III
November, 2012

**INTRODUCTION TO THE SECOND EDITION:**

The second edition of *The Pipeline: A Picture of Homebuilding Production* happened because builders attending our Pipeline workshops™ suggested ways to make the Pipeline game™ better reflect what actually occurs in homebuilding production.

Earlier versions of the Pipeline game™ more resembled a manufacturing operation, in which internal resources determine the capacity and the cost of the system. In this new version of the Pipeline game™, external resources still determine capacity, but they no longer determine the cost of that capacity. The cost of external resources is now a variable cost associated with Cost of Sales (stipulated as a percentage of Revenue), and the capacity cost of internal resources is now an imposed (budgeted) non-variable cost associated with the company's Operating Expense.

Changing the game and restating the operating statement to reflect those changes makes the Pipeline game™ a much more insightful simulation of homebuilding production and a business game that better reflects business outcomes.

Just call it an example of continuous improvement.


**INTRODUCTION:**

*The Pipeline: A Picture of Homebuilding Production* is about the principles and disciplines of production management, as they relate to – and as they are applied toward – the specific conditions, requirements, and parameters found in the homebuilding industry.

These principles and disciplines apply to the production operation of every homebuilding enterprise, but they are most applicable to what we term production homebuilding.

I am told that the Introductions written for business books typically answer two questions: Why should you purchase this book? What is the best way to use it – how do we want you to use it?


*Why should you purchase this book?* In my opinion, you should buy this book, because improving performance on the velocity side of the ROA equation is the best path – perhaps the only path – to achieving sustainable competitive separation.

The issue is not that the margin side of ROA is unimportant – or less important – than the velocity side of ROA. Margin is neither unimportant nor less important; dollar-for-dollar, the Gross Income derived from increasing how much you make on each house you build (margin) has the same value as the Gross Income derived from building more houses with a finite and controlled amount of inventory and capacity (velocity).

Nor is it necessarily a choice. We don't usually have to choose between efforts to increase Return on Sales and efforts to increase Asset Turn; margin and velocity are driven by different aspects of the business, and they don't necessarily react to, or adversely affect, each other.

It's not usually a choice. It's about both.

It is, simply, that higher margin – while as desirable, beneficial, and important as higher velocity – is not a strategy for creating a lasting competitive advantage; between higher margin and higher velocity, higher margin is the easier, more common strategy. The same is even more true of the opposite to higher velocity, which is higher capacity. Adding production capacity (and the inventory for it to work on) is a "more-for-more" proposition. It's the easy, well-traveled road; anyone can resort to adding production capacity, but don't expect it to set you apart.

True, sustainable, competitive separation comes from doing what your competition will not – or cannot – do. Like finding ways to become more productive, to "do more with less".

Consider the plight of RB Builders, the mythical homebuilding company portrayed in *The Pipeline*, facing the world at the close of 2007[2], following the end of the halcyon period known as The Age of Homebuilder Entitlement:

*In many ways, RB Builders was a product of that age, just another homebuilding company satisfied with occasionally adopting other builders' "best practices", content to be good, no-better-but-no-worse than the other builders with whom it competed, a building company with a middle-of-the-road approach to delivering the value its homebuyers demanded.*

*The previous 10 years had been good for RB Builders. But, it was becoming a dangerous approach to business, because – as the saying goes – "the only thing in the middle of the road are yellow lines and dead armadillos".*

*It was becoming a homebuilding no-man's land.*

*Locked into an operating model – into organizational structures, management systems, processes, cultures, and employees – that could not deliver extraordinary levels of distinctive value, the company found itself dumped into a teeming mass of homebuilders that all looked the same, sounded the same, and priced the same. Indistinguishable from other builders, and unable to create any type of competitive advantage, RB Builders was trapped and sinking – like a modern-day dinosaur – into the tar pits of average-ness.*

The world doesn't need any more average homebuilding companies; it has enough of them, plenty of them.

*The Pipeline: A Picture of Homebuilding Production* is structured around the series of team sessions used by RB Builder's intrepid, results-based consultant to build an understanding of production management. It reflects the distinct nature of a homebuilding operation. For a homebuilding company, production management is essentially project portfolio management with embedded production processes and surrounding support processes. Which makes a homebuilding company a project management organization (PMO), and, because it has to manage multiple projects, it is really a project portfolio management organization (PPMO).

---

[2]Excerpt from *The Saga of RB Builders*, Fletcher Groves III (2007).

CHAPTER I creates the visual image of homebuilding production as a pipeline – its purpose, what determines its size (work-in-process), capacity (output), and length (cycle time), and its cost (fixed overhead). It explains the relationship between cycle time, work-in-process, and output. Chapter I discusses the ramifications of utilization as a choice between higher productivity and higher capacity, and it poses the implications of growth.

CHAPTER II looks at the terms that describe the three actions that occur operationally with money in a homebuilding operation (Throughput, Inventory, Operating Expense), and connects the key measures of operating performance (cycle time, productivity, inventory turn) to the key measures of business outcome (Net Income, Return on Assets) with those terms. The chapter explains the complementary roles that margin and velocity play as the components of economic return, and talks about flow, efficiency, and effectiveness.

CHAPTER III explains "systems-thinking", the discipline of thinking, focusing, and problem-solving that is the context in which homebuilding production must occur, a discipline that is rooted in cause-and-effect, interdependent relationships, ordered behavior and outcomes, the way things work, the way problems are solved, and the way constraints are managed.

CHAPTER IV explains the nuance between systems, processes, value streams, and projects, from a production perspective.

CHAPTER V takes up the discussion of production from the standpoint of how a process deals with variation and uncertainty, and how a process is scheduled. The chapter deals with human behavioral tendencies and the manner in which a production system or process protects (buffers) itself from variation. It talks about the differences and similarities between how proven production methodologies, like Lean and TOC, schedule their processes – pace, types of flow, push v. pull production, balanced capacity v. unbalanced capacity, and buffering.

CHAPTER VI discusses the true nature of homebuilding production as project portfolio management, and re-emphasizes the relationship of differences between process management and project management. The chapter explains the main differences (buffering and resource contention) between the critical path method and critical chain project management, and discusses the important changes that must occur in scheduling, in order for a homebuilding operation to reduce cycle time, and to increase productivity and output.

CHAPTER VII introduces the Pipeline game™, a simple, probability-based, team-oriented production management simulation that uses actual game results to sequence the learning and comprehension of principles and disciplines of homebuilding production in a world of variation and uncertainty, and to tie those principles and disciplines together in an effective business framework.

CHAPTER VIII summarizes the contents of the book – the pipeline, the connection between operating performance and business outcomes, systems-thinking, production systems, managing processes, and managing a portfolio of jobs; describes the elements of the production management system upon which production principles and disciplines must act; and, offers a strategic framework in which *The Pipeline* must operate: the *discipline* of a narrow strategic focus;  the *context* of an underlying business logic;  and a horizontal *perspective* of workflow and delivered value.

Chapter VIII concludes with distinguished insight on dealing with dire circumstance:  Ronald Reagan on "not being afraid to see what you see";  James Stockdale (Vice Admiral, USN, Medal of Honor recipient) on the paradox of confronting the brutal facts of a current situation, without ever losing faith in the ultimate outcome;  Merom Klein and Rod Napier on the candor, purpose, will, rigor, and risk required to find the courage to stay in business and build a successful enterprise.

CHAPTER IX explains a revised, improved version of the Pipeline game™ that reflects the out-sourced environment in which homebuilding production occurs, and restates the operating statement to reflect how both direct variable costs and indirect non-variable costs behave in relation to changes in the Revenue that drives them.  It is a production simulation and business game that more reflects a homebuilding operation.

Regarding the second question, *What is the best way to use this book – how do we want you to use it?*:  Personally, I would like to see *The Pipeline:  A Picture of Homebuilding Production* used just as it was intended, as a workbook:  Highlight it, underline it, write in it, tab it, dog-ear it.  Play the Pipeline game™ (see Chapters VII and IX).  Challenge its assertions.  Ask the questions that will give you clarity.  Record your insights about how you would apply the principles and disciplines of homebuilding production to your own situation.

And, then – do something with what you learned.

Enjoy it.

– Fletcher Groves, III

**PROLOGUE TO A PIPELINE:** "The Saga of RB Builders"

As this story is being told, it is the end of 2007, and RB Builders is embarking on a long journey (previously recounted in *The Saga of RB Builders*) to radically improve operating performance and the resulting business outcomes; at the end of 2007, the beginning of 2008, the company could not have known the depth and duration of the housing and economic recession they had entered. Their intent was to structure first one, then another, then several more projects, each with short timeframes and targeted results, each the logical successive step in the pursuit of its overall goal, each the next step in pursuit of its quest for continuous improvement.

This plan to achieve targeted increases in a single business outcome (in RB Builders' case, Gross Income above a currently-achievable baseline) by driving continuous improvement in operating performance through a series of short duration initiatives with targeted, measurable results was complemented by a team-based performance compensation plan that gave every teammate a financial stake in the achievement of that business outcome.

Very early in this effort, the company had concluded (with the help of its intrepid, results-based consultant/partner) that – among its other, not-so-insignificant problems, and despite its considerable experience and past success – it actually knew surprisingly little about the principles and disciplines that relate to homebuilding production. Moreover, RB Builders really didn't have a picture of what production should look like.

In the past, RB Builders tended to sell as many homes as it could, start them whenever it wanted, and finish them whenever it could. In the company's collective mindset, production was the sum of a thousand independent decisions, made without regard for production as a system subject to – and affected by – events of dependency or relationships of cause-and-effect. However, from its new-found perspective of current reality and systems-thinking, RB Builders was now beginning to see the consequence of its production planning and management.

From a production standpoint, the company had always endured long cycle times (upwards of 180 days), low inventory turns, and an uneven rate of sales, starts, and closings. In the final, halcyon years of "The Age of Homebuilder Entitlement", closing dates came and went, while RB Builders' sales managers gleefully spoke of six month "contract backlogs", as if that were some kind of virtue. The contract backlogs were now a thing of the past, but, strangely, the other consequences of RB Builders' production practices remained.

The internal (production) constraint of previous years had been replaced with an ominous external (market) constraint. Still, as in the past, its trade partners complained about jobs that weren't ready as promised, all the while being tugged in different directions, as the company's superintendents (focused on protecting their individual bonuses) fought for resource availability.

In 2007, RB Builders had 200 closings and an average work-in-process of almost 100 homes; the company's construction lines of credit – still larger than its owners preferred – were usually fully-drawn. In 2005, RB Builders had closed 225 homes with the same average work-in-process. The GI Baseline for 2008 was based on the same number of closings as 2007.

In the aftermath of the results-based planning that had preceded it, RB Builders' owners made it clear that while the 2008 GI Baseline and GI Target were based on the company finding ways to do more with what it already had, it was their preference to utilize the existing investment in capacity, not reduce it.

# CHAPTER I:  THE PIPELINE

The intrepid, results-based consultant looked at her watch, stood, and walked to the front of the conference room, now filling with a cross-section of teammates and leaders responsible for production.

On the erasable board, she started a list:

*Pipeline*

"Okay.  We have a lot of work to do", she said.  "Over the course of the next few days, we are going to learn the principles and disciplines that govern homebuilding production.  We need to create a visual reference, and I think the clearest picture – the best visual image – we can convey of RB Builders' production system is that of a pipeline.

"So – going with that image – what is the purpose of this pipeline?"

"To generate income!", someone said.  "To make money.  And – to provide jobs for pipeline workers!"

"Yes, indirectly, as an outcome", she said.  "In the end, RB Builders' goal is to make money from selling and building homes.  Which is not the same thing as its dream, its passion, its purpose, or its core values.  The goal of "making money" is a prerequisite, simply one measure of RB Builders' success.

"But – back to the pipeline.  What does it do?  What is its purpose, what does it carry, and what does it deliver?"

"I would say that the pipeline does two things", offered the VP of Construction.  "It carries our work-in-process – it carries houses under construction – and it delivers closings – completed homes.  So – its purpose is to produce completed homes, and generate Revenue from the closings that ensue."

The intrepid, results-based consultant added to her list.

*Pipeline*
*Pipeline Size v. Pipeline Capacity*
*Cycle Time*
*Work-in-Process*
*Throughput*

"Okay. Then what is the capacity of the pipeline to do that? How many houses – how much work-in-process – can the pipe carry?", she asked, "How do houses get into the pipeline? And – how many closings is it supposed to produce?"

"As many as we can put in it. However we want to put them in it. Whenever we want to put them in", one of the superintendents quipped. "Okay – seriously. We're told we're supposed to generate an even and sufficient rate of sales, starts, and closings.

"That part makes sense. We just can't seem to achieve it.

"And – if we could smooth-out our rate of sales, starts, and closings – then we could probably also manage to maintain a consistent level of work-in-process in the system.

"That part makes sense, too.

"But – as for the capacity of the pipe – apparently we think it has unlimited capacity, because every start we put in the pipeline will eventually be completed and closed. As for the output – the throughput, or whatever you call it – that's a budgeted number of completed houses that turn into closings every year, sometimes we make it, sometimes we don't.

"As far as how houses actually get into the pipeline, there is a start matrix, which acts as the pipe's control valve. Under the old production system, the start matrix prescribed both the order and rate of starts, and 'pushed' the starts into the system. Under the new production system, the start matrix only prescribes the order; houses are supposed to be 'pulled' into the system at the rate of closings."

"More-or-less", said the intrepid, results-based consultant. "Let's go on. I have several questions. First – is there a difference between the size of the pipe and its capacity? Second, how many homes should you have under construction – how much WIP do you need – in order to reach your budgeted closings? Third – how long is it supposed to take you to build a house, and how long does it actually take you? Finally – can you express your budgeted closings as a periodic rate?"

"Regarding your first question – yes – I suppose there is a difference between size and capacity", said the VP of Construction. "The size of the pipeline would be defined by the amount of work-in-process, while the capacity of the pipeline would be a function of output in relation to size. There is a limit to how much it can hold, so – again, yes – the size of the pipe is finite.

"Regarding your second question – again, yes – there is a connection between how much the pipe can hold and how much it can produce. We think the pipe should be able to hold 100 houses, and we think the pipeline should be able to produce 240 completed houses a year – at least, that's the budget – which is 20 closings per month. So – you could say that the size of our pipeline is 100 houses, and its capacity is 20 completed houses per month.

"In terms of our cycle time, it varies slightly depending on the house plan, but our construction schedules call for us to average 120 days", the VP of Construction continued. "However, we know we are nowhere near that fast.  Most of our homes finish late.  I would say that eighty (80%) percent of our houses take between 160 and 200 days."

The intrepid, results-based consultant thought for a moment about what the VP of Construction had said.  "We are going to presume that all of these completed houses become closed homes, so we will say those terms are synonymous, even though we know there is some additional time;  it's not a production issue, and it helps with the principles you are going to learn.

"Now, if you closed 240 homes with 100 houses in WIP, your cycle time would be about 150 days", she said, "30 days longer than expected.  However – this year – you are only on track to close 200 homes, which means your cycle time is pushing 180 days."

"Then we agree", replied the VP of Construction, addressing everyone.  "The way we measure it, our cycle time has been averaging around 180 days.  There is considerable variation, particularly on individual jobs;  some take more time, some take less time.  But, the overall average is around 180 days."

The intrepid, results-based consultant moved to the erasable board at the front of the conference room, selected an erasable marker, and wrote:

$CT = 120$ days   $WIP = 80$ houses   $T = 240$ homes

$CT = (WIP \div T) \times$ Days
$(80 \div 240) \times 360 = 120$ days

$WIP = (CT \times T) \div$ Days
$(120 \times 240) \div 360 = 80$ houses under construction

$T = (WIP \div CT) \times$ Days
$(80 \div 120) \times 360 = 240$ closings

"There is, in fact, a direct connection", she said.  "There is an accepted, proven mathematical relationship between the length of process cycle time (CT), the level of work-in-process (WIP), and the throughput (T) – or the output – of a process, expressed as a periodic rate. So – if you know two of the values, you can always calculate the third value.

"There are two laws of production that deal with this relationship.  The first one that I just mentioned, which is called Little's Law, and a second law, one which we call the Law of Variability Buffering, which tells us that every system will protect itself from unplanned variation and uncertainty with some combination of – you guessed it – longer cycle time, more inventory (work-in-process), or excess/unused capacity."

The intrepid, results-based consultant wrote:

## Systems

"All of which points to the fact that we live in a world of systems.

"The homebuilding industry, the housing and real estate market, and the local and national economies in which a homebuilding company operates – they are all part of a system. The business environment within which a homebuilding company must operate is also a system. These production principles and disciplines are part of a system.

"A homebuilding company is not some loosely-connected set of independent, unrelated parts – a loose collection of processes, departments, systems, resources, policies, and other isolated pieces of a whole. A homebuilding company is both a *system*, and a part of a system – a set of interdependent parts that must work together to accomplish a stated purpose.

"Viewed as a pipeline, production systems have neither unlimited capacity nor unlimited size.

"If you increase (the level of) work-in-process, the only way the system can hold the additional work is to lengthen the pipe. The diameter of the pipe is fixed; if we put more work-in-process in the pipe, it doesn't become a bigger, wider pipe – it just becomes a longer pipe. So – what is the length of the pipe?"

"The length of the pipe is the duration to build a house. It's cycle time", replied the VP of Construction.

"That's right", she said. "Duration – or cycle time – is the measure of the length of the pipe. The longer the pipe, the more time it takes to get from one end of it to the other. In fact, given the same amount of effort, the friction, the increased number of turns, etc., resulting from the added length, actually tends to reduce the output."

A superintendent raised his hand. "Okay. So – are you saying we need a bigger, wider pipe?"

The intrepid, results-based consultant quietly smiled. "Well, that depends", she replied. "Does your production pipeline have a cost?"

"Everything has a cost", said the VP of Construction, turning to the CFO. "Isn't that right?" The CFO smiled wryly, nodded affirmatively, and replied, "Yes – it does."

"So – what is the cost of your pipeline?", she asked, adding to her list.

## Pipeline Cost

"Well, we've never thought about it that way", the CFO responded. "I suppose the cost would be whatever we spend to have a pipeline in place. It seems to me that the nature of a production pipeline is that of a relatively fixed object – you know, heavy and difficult to move. I know I wouldn't want to move it. I would say that the cost of our pipeline is all of the expenses we incur every year, to have the capacity to build houses."

"That's right", she replied. "The cost of the pipeline is what we pay every year, in the form of operating costs and resources, to have the use of it. We pay for the cost of the pipeline, whether we use it or not. That puts the cost of the pipeline squarely in the category of non-variable costs.

"Which brings up another point. To understand productivity and production capacity, you must first understand how costs behave (in relation to Revenue), and how you manage those costs on the basis of that behavior.

"On the one hand, you want to *control* your direct, variable costs – meaning you want to reduce the cost. Really, though, what you want to do is extract maximum value from it. Value is the difference between the price you sell a house for, and what it cost you to deliver it. On the other hand, you want to *leverage* your indirect, non-variable costs; those are the costs you expect to incur regardless of the Revenue you generate, and you want to produce as much output (Revenue, resulting in Gross Income) as you can from them.

"So – would a bigger, wider pipe cost more than your current pipe?"

Thinking for a moment, RB's CFO replied, "Yes, it would. There is a connection between the size of a pipe and its cost. There is also a connection between the size of a pipe and its capacity, but that's an issue of utilization. When we invest in a pipe, the cost of the pipe is based on its size.

"So – yes – a bigger, wider pipe would cost more than our current pipe."

"You mentioned utilization", said the VP of Construction. "Our production pipeline is almost always full. So – are you saying that we don't utilize our production capacity?"

"No. Well – maybe", said the CFO. "I don't know how effectively or efficiently we are using the capacity that the pipe was designed to achieve. All I'm saying is that there is a relationship between the size of the pipe we design or buy, and what it costs us. The price of the pipe is related to its size, and that cost is fixed. It's up to us to utilize the investment, to use the capacity."

"That's right. I want to summarize the definitions of all these terms. A pipeline's *size* is defined by the amount of work-in-process it is intended – or designed – to carry", said the intrepid, results-based consultant. "Its *cost* is its Operating Expense, which tends to be non-variable. Its *length* is its cycle time. Its *capacity* is defined as the rate of output – or throughput – a pipeline that size can produce, with a planned, finite, and controlled level (or amount) of work-in-process."

Writing on the board, she continued, "You can turn the definition of capacity around, and look at *capacity* as the level of work-in-process required to support a targeted rate of throughput, but, essentially, it's the same thing."

Size = Work-in-Process
Length = Cycle Time

*Capacity = Closings with a controlled level of WIP*
*Control = Rate of Closings and capacity of the scheduling resource*
*Cost = Operating Expense and Resources*

"Earlier, someone described a *control valve* that allows starts into the pipe", she said. "Actually, there are two control valves. The first control valve is the rate of closings. To our way of thinking, it is located at the end of the pipe. The second control valve is located inside the pipe, and it is the production rate of the constraint resource that schedules all of the other resources. When we say that it is a 'constraint resource', we mean that it is the resource with the least amount of capacity, relative to the demand that is being placed on it.

"The first valve – representing the rate of closings – makes the starts available, and controls the level of work-in-process. The second valve – representing the drum resource or pacemaker – pulls the starts into the system; that constraint resource is formally known as the Capacity Constraint Resource. So – the proposition is not exactly 'close one, start one', it's more like 'close one, start one, as soon as the CCR says you can'.

"The two control valves together enable us to have a pull system, and, ideally, they are synchronized, so that we can, in fact, 'close one, start one'. This staggering or pacing of jobs is known as 'pipelining' in a production system that has to manage multiple projects (or jobs).

"The valves are how you manage the pipeline as a system.

"Lastly – the *cost* of the pipeline is what we pay each and every year – the indirect, non-variable operating costs, and all of the resources associated with those costs – to have the use of it."

"I'm confused", said the first superintendent. "Are we – or, are we not – fully-utilizing the capacity of our production pipeline?"

"I don't think we are", replied a second superintendent. "We've already said the pipeline was designed to produce 240 closings a year with 100 houses in work-in-process. We have the inventory, but we don't have the throughput. Plus, our current cycle time is 180 days, not the intended 150 days, certainly not the 120 days called for by our build schedule."

"We can talk about the effect of long cycle times and why that has happened later", said the intrepid, results-based consultant. "But – I agree – you have not been fully-utilizing the capacity you have been paying to have. We can talk about why that has happened later, too."

She continued the list.

*Size and Growth*
*Adding Production Capacity v. Increasing Productivity*

"Right now – let me ask a different question: What were your choices for dealing with the issue of capacity utilization? Before the current downturn in the housing market – something else we can talk about later – what would your alternatives have been for getting throughput up to the designed level?"

"Based on what I've learned from this discussion, I would say we probably had two options", said the VP of Construction. "We could have added production capacity; technically, that shouldn't really count as an option for increasing utilization, because it alters the designed capacity of the pipe.

"The other option would have been to better utilize the production capacity we already had."

"What is the operational term we use for option two?", she asked.

"We would be increasing productivity?" suggested the second superintendent.

"Exactly", she said. "There are only two choices. You can either add capacity or become more productive. It is a decision that cuts to the core of how you view size and growth. What size is RB Builders?"

"$50 million", answered the VP of Sales. "That was our Revenue for 2007."

"And, that is how most homebuilders would answer the question", she said. "The answer to the question of size is usually about the amount of annual Revenue or the annual number of closings. However, the most relevant measure of the size of a homebuilding company is the amount – and the value – of the work-in-process that it carries.

"Size is about capacity, not output.

"As we have already seen, there is a direct correlation between work-in-process and production capacity, which we prefer to define as the rate of throughput (or output) that can be generated with a planned, finite, controlled level of work-in-process. There is an equally strong and direct correlation between work-in-process, cycle time, and velocity (or Inventory Turn). Finally, work-in-process is one of the ways a production system will protect itself – buffer itself – from variation and uncertainty."

"You mentioned growth along with size", the CFO reminded her.

"Yes, I did", she said. "If size is defined as capacity, rather than Revenue or closings, what is the implication for growth? How should RB Builders grow?" Noting the blank stares all around the room, she continued. "The answer, based on that definition, is that RB Builders should *not* want to grow.

"By that definition, even when faced with acceptable justification, RB Builders should see growth as a last resort. RB Builders doesn't want to add production capacity, it wants to increase its productivity, by increasing the utilization of the production capacity it already owns.

"Adding production capacity – getting bigger – is a 'more-for-more' proposition", she continued. "It's the easy road. Anyone can do it. Anyone can resort to adding production capacity, resort to spending more money.  True, sustainable competitive separation comes from doing what your competition will not – or cannot – do.  Like finding ways to become more productive.

"Beyond the competitive aspect, there are other problems that come from simply being big", she said.  "Big homebuilding companies tend to be slow, clumsy homebuilding companies, unable to respond quickly to changing circumstances, incapable of exploiting opportunities in the marketplace.

"And – there is risk," she added.  "Adding production capacity means additional work-in-process and additional resources.  Risk increases exponentially with an increase in core size – with higher WIP and Operating Expense.  Once you increase production capacity, it becomes much harder to fully utilize it.  There are fewer options.  It is very difficult to downsize your way out of excess production capacity.  Size forces you into positions you shouldn't be in;  size forces you down roads where you shouldn't go.

"Okay – so let's talk about the other option," she said.  "What do we mean by the term 'productivity'?  How do you increase productivity?  What does it mean to become more productive?"

"Isn't that the same question?", asked the CEO.  "Or – is it a different question asked the same way?"

"The question was about productivity", she said, ignoring the interruption.  "Any thoughts?"

"If adding production capacity is a 'more-for-more' proposition, then I suppose improving productivity would be a 'more-for-less' proposition", said the second superintendent.  "Or, at least, a 'more-for-the-same' proposition."

"Not bad", she said.  "So – what does this 'more-for-less' idea look like?  How do you measure productivity?"

Turning to the CEO, she smiled and said, "Don't wear yourself out."

The CEO smiled and replied, "Productivity is the relationship between what is produced and what has to be consumed in order to produce it."

"That's right", she said, continuing the list.

## Productivity Measures

"From any managerial standpoint – operations, manufacturing, production, or otherwise;  from any industry standpoint – auto manufacturing, homebuilding, or any other industry;  from any enterprise standpoint – Toyota, RB Builders, or anyone else;  from any expert or business leader standpoint – Peter Drucker to Eli Goldratt to Taiichi Ohno, the conventional, accepted formula for calculating Productivity is Revenue divided by Operating Expense."

She moved back to the erasable board, and wrote:

Productivity = Revenue ÷ Operating Expense
Productivity = Output ÷ Input

"Less commonly, you will also find productivity expressed as the ratio between the 'input' and the 'output' of a process", she said. "Either one will do, but the first formula fits best with the correct understanding of what it means to "make money". We will talk about that later."

She continued, "Under either formula, productivity is about what is produced and what is consumed. The 'what is produced' part is pretty clear; we understand what is meant by 'output'. What about 'input'? What is it that is consumed? Is it an asset, or is it a resource?"

"Assets are converted or transformed. Resources are consumed", said the CFO. "Inputs are expenses, just like the first formula."

"That's right", said the intrepid, results-based consultant. "It is an expense. But – what type of expense is it? Is input a fixed cost – like Operating Expense or overhead – or, is it a variable cost?" Turning, she wrote:

## Variable Costing

The CFO stood. "Earlier, you used the terms 'direct, variable cost' and 'indirect, non-variable cost' to describe cost behavior", he said. "I could quibble that direct/indirect and variable/non-variable refer to different characteristics dealing with objects and behavior, but – I agree – that ties with the idea that consumption of a resource would make input an indirect, non-variable cost, as opposed to a direct, variable cost, which is really more like the contra-asset associated with our work-in-process. Like a lot of other homebuilding companies, RB Builders has not clearly separated those costs, but there are certainly advantages to variable costing. In fact, those direct, variable costs don't even become expenses until after we close the job out."

"I agree", said the CEO.

"What about us?", asked the second superintendent. "The argument can be made that the cost of a superintendent is a direct cost, but clearly not a variable cost. The same could be said about construction interest, albeit for different reasons. So – what are we?"

"You are an incredibly valuable resource that happens to be a non-variable cost", said the CFO. "But – you", he said, looking at the first superintendent and grinning. "You are a totally worthless, soon-to-be-eliminated-thus-no-longer-non-variable cost."

"I agree", said the CEO.

The intrepid, results-based consultant put the erasable marker down, and waited until the laughter died down and she again had everyone's attention. "We have talked – briefly, and at different times – about how costs are classified", she said. "We mentioned it in the discussion about the cost of the pipeline. It is part of managerial accounting, more a part of the business principles and disciplines you will be learning than the production principles and disciplines we have been learning. However, we will address variable costing further down the road, because it impacts so many areas of management.

"For now, it will be enough for you to just remember this: In order to understand productivity and production capacity, you must understand how costs behave, and how you manage those costs on the basis of that behavior. Control your direct costs, because they vary in accordance with Revenue. Leverage your indirect costs, because they are more-or-less fixed, and you incur them regardless of how well you utilize those resources.

"There is more to cover, but we have already covered a lot", she said. "It is a very long road. We will get to the rest of it in due time. But – this is what we have covered so far."

Moving back to the erasable board, she walked the team back through each of the points she had previously listed.

### Pipeline

"We said the best description of a production system was a pipeline, and we want to be clear on its *purpose*, on its *size*, its *capacity*, and its *cost*."

### Pipeline Size v. Pipeline Capacity

"We said the size and capacity of a pipeline are not the same thing. *Size* is defined by the amount of work-in-process it carries. *Capacity* is the rate of throughput (or output) in relation to the size of the pipe."

### Cycle Time
### Work-in-Process
### Throughput

"We saw that there are three measures – the length of cycle time, the level of work-in-process, and rate of output (or throughput) – that are crucial to our understanding of production and how production systems are managed. These measures are connected to each other – each affects the other two."

### Systems

"We said that we live in a world of systems – a set of interdependent parts that must work together to accomplish a stated purpose. Systems-thinking is not sum-of-the-parts thinking. Our beliefs about systems has tremendous implications for everything we do."

### Pipeline Cost

"We learned that the *cost* of the pipeline is what we pay every year, in the form of operating costs and resources, to have the use of it. We pay for the cost of the pipeline – for the cost of the capacity – whether we use it or not. We own the pipeline.

"And – right now – RB Builders is not utilizing its investment in its production capacity the way it should."

*Size and Growth*
*Adding Production Capacity v. Increasing Productivity*
*Productivity Measures*

"We said that the choice of whether to add production capacity or improve productivity cuts to the core of how we choose to view size and growth. We don't want to become larger, by adding production capacity. Instead, we want to become more productive, by removing waste and variation, by making our production flow.

"We want a proposition that is about 'more-for-less', not one that is about 'more-for-more'.

"We saw the risk of simply becoming bigger, instead of embracing the discipline of becoming more productive. And – we learned what productivity is, and how it is measured. Productivity is about what is produced and what is consumed in order to produce it."

*Variable Costing*

"Finally – we learned the importance of understanding how costs behave in relationship to changes in Revenue, and how we manage those costs on the basis of that behavior. Variable costing is really managerial accounting, and it is discussed in more detail as part of RB Builders' business principles and disciplines.

"But variable costing is also central to our understanding of production capacity and productivity. Some costs – our direct, truly-variable costs – we need to control and extract maximum value from; other costs – our indirect, non-variable costs – we need to leverage."

The intrepid, results-based consultant capped the erasable marker, set it down, and moved closer to the group.

"This was a good start", she said. "Think about what we have learned, and begin to find ways to apply what you have learned in your decision-making. Next session, we will pick up where we ended, and begin to connect key measures of operating performance to key business outcomes, from the standpoint of production.

"You did great. You were a very attentive, very engaged group. I appreciate your efforts."

# CHAPTER II: THE CONNECTION

The intrepid, results-based consultant finished her catered-in-the-conference-room buffet-style breakfast (two scrambled eggs medium, two strips of hotel-style honey-maple bacon, spicy corned beef hash, breakfast potatoes, buttered Cuban toast with fresh guava preserves, fresh-squeezed orange juice from a remote grove in Highlands County, black coffee, and a palate-freshening dollop of grapefruit sorbet), and moved to the front of the conference room.

"In our last meeting, we talked about the image of production as a pipeline", she said. "We also acknowledged, that while 'making money' is RB Builders' goal, it is not the same as its purpose, passion, dream, or measures of success. So, today, we want to begin to connect the business outcomes reflected in that goal to the operating performance that drives it.

"We talked about this connection in the context of the business principles and disciplines we have been learning elsewhere, but it is also relevant – essential, really – to a proper understanding of production principles and disciplines, so", she continued, "someone please remind us 'what happens to money' in a homebuilding company, from an operational perspective."

"We receive money from closings, we use some of our money to build the houses, and we use some of our money to pay our bills", someone said. "Oh yeah – and the owners keep some of it".

"Or lose some of it. The owners haven't had much to keep lately", said the CEO.

"More-or-less", said the intrepid, results-based consultant, responding to the first answer. "As part of the new focus on results, RB Builders' owners are giving everyone a financial stake in achieving an improved level of operating performance and business outcomes. However – team-based performance compensation is not what we're here to talk about right now.

"RB Builders generates money from closings", she said. "How much money?"

"In 2007, we closed 200 homes for $50 million. So – our average sales price is about $250,000. That's all money we generate from closings", said the VP of Sales. "That's also our baseline for next year."

"But, we don't get to keep all of that money", said the CFO. "If you look at the HUD-1, it might look like that's what we get to keep, but that's a timing situation, and – not with us, but with some builders – it can also be a financing situation. In any event, we only get to keep the Revenue generated by the closings, less the truly-variable costs associated with the land, the cost of building the house, perhaps the construction period financing, and the selling and closing costs incurred in the sale. So – in 2007 – we kept $11 million of the $50 million in Revenue."

"That's right", the intrepid, results-based consultant said. "That is what is known as *Throughput*. It is not the same as the throughput we said is synonymous with 'output', which is the physical (or operational) meaning of the term. In monetary terms, Throughput is more like Gross Income, or, perhaps, Contribution Margin.

"Secondly, the money we used – temporarily – to pay for the land and build the house is known as *Inventory*; it's really the monetary value of our work-in-process", she continued. "So – now we have Throughput and Inventory. You mentioned a third thing that 'happens to money', something you referred to as money we 'use to pay our bills'. What do you mean?"

"That's our overhead", said the CFO. "The last time we met, we discussed how you classify costs according to behavior to the changes in Revenue – a variable costing approach – and we said that our truly-variable, direct costs are our Cost of Goods Sold, or Cost of Sales, what we sometimes call COGS or COS. These are costs that appear 'above the line' on our Income Statement, but – before closing – they appear as an asset on the Balance Sheet. This is what we are referring to as Inventory; these are the costs associated with what we 'convert' or 'transform' into output. Earlier, someone correctly noted that it's the money we use to 'build the houses'.

"Okay", he continued. "The costs we use to 'pay bills' is different. Those are our non-variable, indirect costs; we say they are 'consumed' costs. We talked about these non-variable, indirect costs as representing the cost of our production capacity.

"It all gets down to how these costs behave in relation to what drives them, which is changes in Revenue. We want to control – to either reduce or extract value from – our COGS, but we want to leverage our overhead – to effectively and efficiently utilize the cost of our production capacity."

The intrepid, results-based consultant nodded her head in agreement, and said, "These non-variable, indirect costs are known as *Operating Expense*. In 2007, RB Builders had $8.5 million in Operating Expense, resulting in Net Income of $2.5 million."

She moved to the erasable board, selected a marker, and wrote:

Throughput
Inventory
Operating Expense

"So – these are the three monetary occurrences. These are the three 'things' – as a matter of fact, the only things – that can 'happen to money' from an operational point-of-view", she explained. "*Throughput*, or Gross Income[3] is all of the money a building company *generates*

---

[3] Throughput represents the proceeds that a homebuilding company gets to keep from every closing. That makes Throughput more synonymous with Gross Income (or Contribution Margin) than with Revenue. For

through closings. *Inventory*, or work-in-process[4] is all of the money a building company *invests* in what it intends to turn, or more accurately, transform, into Throughput. *Operating Expense* is everything a building company *spends* to transform its Inventory into Throughput."

On the erasable board, she wrote:

T - I - E = two sides of the same coin

"Like two sides of the same coin, financial operating measures have correlating physical operating measures. *Throughput is the rate of closings, Inventory is the number of houses under construction*, and *Operating Expense is expressed as the equipment and resources tied to its non-variable costs.*

"This part is important", said the intrepid, results-based consultant, waiting for everyone's attention. "As terms of operating performance, Throughput, Inventory, and Operating Expense – closings, work-in-process, and resources – move in response to every operating decision RB Builders makes.

"Unlike financial performance, which is an outcome, operating decisions are happening right now, every day, and we have control over them. We have to figure out how to connect the work we do to the outcomes we want. *In other words, we have to be able to link operating performance to financial performance, through the terms operating and financial performance hold in common: Throughput, Inventory, and Operating Expense.*"

She moved back to the erasable board, and wrote:

Cycle Time = Inventory ÷ Throughput
Productivity = Throughput ÷ Operating Expense
Inventory Turn = Throughput ÷ Inventory

"These are the three operating measures – the performance drivers – we want to link back to business outcomes", she said. "The first two formulas – cycle time and productivity – should be familiar, because we discussed both of them during our last session. The third measure – Inventory Turn – we might have touched on last time, but I have not given you the formula.

"As you can see, what the formulas for these three operating measures have in common", she said, "are their variables: Throughput, Inventory, and Operating Expense. And – with a

---

some purposes, there are calculations (for example, Inventory Turn or Return on Assets) where Revenue is more acceptable, but the understanding of Throughput as 'money generated' remains true.

[4] Inventory is synonymous with work-in-process, but it also includes a homebuilding company's investment in vacant lots and undeveloped land held on its Balance Sheet. The vast majority of the "Inventory" (or "Investment") a building company carries is in the form of work-in-process (or WIP), reflected in the loans-in-process (LIP) balances on its construction lines of credit or individual construction loans.

couple of minor caveats[5] – the way we link these three operating measures to the two business outcomes we are focused upon, is through these variables."

Turning back to the erasable board, the intrepid, results-based consultant wrote:

Net Income
Return on Assets

"These are the two business outcomes we want to continuously improve. Plus, of course, Cash Generation, or cash flow", she said. "So – someone give me the formula for Net Income, but using only the terms Throughput, Inventory, or Operating Expense."

"Throughput is Revenue, less all of the truly-variable, direct costs, which makes it the equivalent of Gross Income", said the CFO. "Net Income is calculated as Gross Income, less all of the non-variable, indirect costs generally referred to as overhead. So – using only T, I, or E, the formula for Net Income would be Throughput, less Operating Expense."

Moving to the erasable board and asking the intrepid, results-based consultant for her marker, the CFO completed the Net Income formula:

Net Income = Throughput – Operating Expense

"And, now that we have Net Income – if I know where you are headed with this – the formula for Return on Assets should be something like Net Income divided by Inventory", the CFO said, and then wrote:

Return on Assets = (Throughput – Operating Expense) ÷ Inventory

"However, we use the expanded, more informative formula for Return on Assets – or Return on Invested Assets – called the DuPont Formula, so that we don't lose sight of the margin and velocity components of economic return.

"The DuPont Formula is the product of two quotients. The first quotient is from what we call Return on Sales, which is Net Income divided by Revenue, and the second quotient is what we call Asset Turnover, which is Revenue divided by Average Assets.

"In a building company, most of the assets are in work-in-process, or inventory of some sort. It doesn't have a lot of fixed capital cost or investment, like plant and equipment. So, the term Inventory works. There's only one problem. According to the rules, we can't use Revenue. We can only use T, I, and E, which is okay, because Revenue cancels out when you solve the DuPont Formula. Net Income is the remainder of Throughput minus Operating Expense, so the shorter, more common formula presents no problem."

---

[5] In order to calculate cycle time for a specific period of time – monthly, quarterly, semi-annually, annually, etc. – you have to prescribe the number of days in that period, which we say are 30, 90, 180, and 360 days, respectively. In order to calculate any measure involving assets (which is every operating measure and business outcome involving Inventory), you have to use the average assets for the period; you can compute a daily average, but for most purposes, the average of the beginning and ending of the period is fine.

The CFO wrote:

$$\text{Return on Assets} = (\text{Net Income} \div \text{Revenue}) \times (\text{Revenue} \div \text{Inventory})$$
$$\text{Return on Assets} = \text{Net Income} \div \text{Inventory}$$

"That's true. But – remember – we are primarily looking at operating measures and linking them to business outcomes", said the intrepid, results-based consultant. "When we substitute the physical operating measures for the financial measures – when we flip the coin, so to speak – the relationships becomes more clear.

"Look at the operating measures", she said. "Cycle time is a function of the rate of closings and the amount of work-in-process, productivity is a function of that same rate of closings and the resources required to produce it, and Inventory Turn – as a function of the level of work-in-process and the rate of closings – is the reciprocal of cycle time.

"Think about it. In 2007, RB Builders had a cycle time of 180 days, a function of an annual rate of 200 closings – its Throughput – produced with a level of 100 houses in work-in-process – its Inventory. Reverse the formula. What was RB Builders' 2007 Inventory Turn?"

"2-x", said the VP of Construction. "During 2007, we turned Inventory twice."

One superintendent turned to another superintendent, and said, "It all just makes you wish you had been an accountant or a consultant, doesn't it?"

"Excuse me. May I say something?"

Everyone's head turned in the direction of the CEO, who stood and walked to the front of the conference room. "Look", he said. "What we want is understanding. We want business literacy. Every important measure of operating performance – Cycle Time, Productivity, Inventory Turn – and every important measure of financial performance – Net Income, Return on Assets – can be calculated using three terms: Throughput, Inventory, and Operating Expense.

"It doesn't bother me that we may need to modify the DuPont Formula. It doesn't bother me that we may need to change our thinking on cost behavior and cost reporting. What I care about is that every Teammate and every Leader understands the connection.

"Because, if we make operating decisions – each day, everyday – that maximize the rate of Throughput, i.e., that maximize the rate of closings, with a planned, finite, and controlled level of Inventory, i.e., with a planned, finite, and controlled level of work-in-process, and we do it by leveraging our investment in our production capacity – if we focus on doing those three things well – then we will have shorter, faster cycle times, higher productivity, and faster turns.

"And – if we have shorter, faster cycle times, higher productivity, and faster turns, then we will be more profitable, we will earn a higher economic return, and we will generate more cash.

"As one of our great, illustrious owners would say, 'Bottom-line. End of story'.

"RB Builders has already taken steps in that direction", continued the CEO. "As part of the new, focused, results-based approach, we have already moved to limit and control the amount of work-in-process and the cost of our production capacity. The owners have stipulated that the improvements in operating performance and business outcomes we are targeting under this new approach will have to reached without any increase in work-in-process or production resources.

"Personally – I believe we should be looking for the opportunity and justification to reduce the level of work-in-process even further, in order to operate as closely as possible to a 'zero working capital' position.

"But – I also understand that we need to take this one step at a time, so that we learn from the outcomes and incorporate that learning into successive projects, so that we can innovate, so that we can continuously increase our capability and capacity to improve. If reducing work-in-process further will make us faster, if fully utilizing our existing capacity will make us more productive, I think we will come to that conclusion over time and take the right steps.

"I also understand that we need to live in the 'Throughput World', not in the 'Cost World'. There is no theoretical limit to the Revenue and Gross Income we can generate, but there are limitations to the amount of capacity – the amount of work-in-process and resources – we can safely eliminate. Max 'T' with a planned, finite, and controlled amount of 'I' and 'E'.

"That's what we want. That's what improves operating performance and business outcomes."

Recognizing and appreciating the personal watershed his statements likely represented, the intrepid, results-base consultant nodded, inconspicuously, yet knowingly, to the CEO, as he returned to his seat.

Then, she wrote:

*Necessary WIP vs. Maximum WIP*
*Necessary WIP vs. Minimum WIP*
*Debt-to-Equity Ratio*

"Okay, let's talk about the amount of work-in-process – the amount of WIP – that we need to have or that we should have", she said. "Recall that we said the size and capacity of a pipeline are not the same thing. *Size* is defined by the amount of work-in-process it carries. *Capacity* is the rate of throughput (or output) in relation to the size of the pipe.

"The question becomes, how much work-in-process do we need or want? The answer is, we want only what we need. We only want what we need to produce the output, we only want what we need to utilize the capacity. We only want what we can use. If we have too much work-in-process, it is size that is wasted, size we paid to have but didn't use. On the other hand, if we don't have enough work-in-process, it's dangerous, because we can't protect the capacity of the pipeline from variation and uncertainty.

"So – how much work-in-process is enough?"

"I will speak to that question", said the CFO. "RB Builders has now adopted a very disciplined business model, so the company pays close attention to its Debt-to-Equity Ratio. D/E prescribes the amount of equity, which in turn limits the amount of debt, primarily used to fund the level of work-in-process; I suppose you could also say the D/E Ratio tells us how much equity is required for a prescribed level of debt.

"That translates into three scenarios of how much work-in-process RB Builders can, should, or must have.

"The first scenario is something we call 'Maximum WIP'. Max WIP translates into a maximum number of units RB Builders is *allowed* to have under construction, and a ceiling on the amount that can be outstanding under the company's construction lines of credit. In the case of work-in-process, more is not better, and Maximum WIP is definitely not where we want to be; just because you can, doesn't mean you should.

"The second scenario is what we call 'Necessary WIP', which is the number of houses – and the associated amount of construction debt – that RB Builders *should* have, given the company's current ability to utilize its production capacity and achieve its targeted Revenue and closings.

"The difference between Max WIP and Necessary WIP is the difference between what is currently needed and what is allowed. Hopefully, what we need is less than what we allow ourselves.

"There is another gap, and that is the gap between Necessary WIP and the third scenario, what we could call 'Minimum WIP'. The measure of the gap between Necessary WIP and Minimum WIP is often the gap between targeted and achieved cycle times, the gap between targeted and achieved levels of productivity, the gap between targeted and achieved utilization of production capacity. Minimum WIP is like, 'Yes, we currently build very fast, but we have to have enough to build, or we can't generate sufficient Throughput'.

"Necessary WIP is based on current cycle time and capacity, so one of the results of an improvement in cycle time, inventory turn, capacity utilization, and productivity is a decreasing level of Necessary WIP, towards Minimum WIP – which is obviously something we want to see."

Moving to the erasable board, the CFO wrote:

Max WIP ---------- Necessary WIP ---------- Minimum WIP
Margin x Velocity

"The amount of work-in-process – size – is connected to the issue of capacity, the issue of productivity, the issue of speed; not just speed, but velocity", the CFO continued. "It's connected to everything we have been talking about. In the end, it is all very connected to business outcomes – to profitability, economic return, and cash generation."

"Dumas", said one of the sales representatives, looking at a superintendent. "Do you know the difference between speed and velocity?"
"I can't see the difference", replied the superintendent.

"Velocity is a vector measure", said the sales representative. "Unlike speed, velocity has direction; it's speed with a purpose. So – when you're running around in circles – you don't have velocity, you just have speed."

The VPs of Sales and Construction looked at each other. The VP of Sales spoke first. "At the break, we were talking about margin and velocity being the two components of the financial measure RB Builders uses for economic return, which is Return on Assets.

"The measure of margin is Return on Sales, while the measure of velocity is Asset Turn. I am focused on margin."

"And, I am focused on velocity", said the VP of Construction. "We would like to know more about the relationship between margin and velocity. How do they work together, and where might they create conflict?"

"Return on Invested Assets is a composite measure of economic return", said the intrepid, results-based consultant. "The formula for ROIA, or the broader measure of ROA, for that matter, is basically margin x velocity. It is a reflection – and an outcome – of both the margin we earn, and the velocity we generate. Return on Sales is the margin part, and Asset Turn is the velocity part; margin and velocity work together to produce economic return. Margin is how much money we make on every home we close, and velocity is about how many houses we can build and close.

"However, while economic return – ROIA – is a composite of both measures, margin and velocity are driven by different aspects of the business.

"Margin is marketing-related. It is a reflection of product development, marketing, and sales; it is a function of upstream and downstream marketing, a reflection of pricing, costs, job budgets and trade partnering, a function of floor plans, elevations, options, specifications, and communities, a reflection of 'buying low and selling high'. From a monetary standpoint – from a managerial cost standpoint – margin is about extracting value from direct, variable costs; it is about exploiting Cost of Sales.

"Categorically, these costs are in each house budget; if you build the house, you incur the cost. if you don't build the house, you don't incur the cost. There's more to it than this, but – basically – if the cost doesn't add or create value, you shouldn't incur the cost.

"Velocity is focused on production. It is a function of job scheduling, production planning, and project management, a reflection of productivity and production capacity, a reflection of the inherent relationship between Inventory and Throughput, a function of 'doing more with less'. From a monetary standpoint – from a managerial cost standpoint – velocity is about leveraging indirect, non-variable costs, leveraging Operating Expense, leveraging overhead.

"Categorically, you're going to incur these costs regardless of how many houses you build, so you need to get more out of it.

"The limitation to higher margins is usually in the market, and we characterize market constraints as being 'external'. On the other hand, the limitation to higher velocity is generally production-related, and we characterize production constraints as being 'internal'.

"Except under dire circumstances, we are not forced to choose between efforts to increase Return on Sales and efforts to increase Asset Turn. As I said, margin and velocity are driven by different aspects of the business, and they don't necessarily react or adversely affect each other.

"It's not a choice. It's about both", she said. "For example, the way RB Builders wants to partner with suppliers and sub-contractors involves both margin and velocity.

"RB Builders and its trade-partners want to find ways to share in the increased value they jointly extract from the houses they build and sell. That's about margin. At the same time, they want to benefit from the shorter cycle times and higher productivity that results from better scheduling and coordination, from allocating production resources more effectively, from finding ways to increasingly do 'more with less'. That's about velocity.

"Both margin and velocity benefit from efforts to wring waste and non-value-adding work out of the system. Both margin and velocity benefit from efforts to eliminate variation and uncertainty.

"It's not about choosing to focus on improving either margin or velocity. It's about choosing to improve *both* margin and velocity", she repeated. "It's a two-pronged attack. Everyone has a dog in this fight.

"In fact, that is the beauty of focusing on Throughput. Throughput increases as a result of both higher margins – more income on every job – and higher velocity – more closings in every community.

"Throughput does not distinguish between Revenue earned by higher margins or Revenue generated with more velocity. Throughput is Throughput. Gross Income is Gross Income. Contribution is Contribution.

"It's about higher margins *and* higher velocity, not higher margins or higher velocity.

"The referee in any supposed dispute between margin and velocity is whatever action increases Throughput.

"If there are any meaningful distinctions, at all, between an emphasis on margin or velocity, they are that (1) higher velocity produces its additional Throughput with fewer assets deployed, and that (2) perhaps we have more control in overcoming internal, production-related constraints than we do in overcoming external, market-related constraints.

"Is that clear?", she asked. "Does it make sense?"

"I want to comment briefly on part of what I said earlier", said the CEO.

Looking at the intrepid, results-based consultant, he said, "I know this isn't what we are here to talk about." Addressing the team, he continued, "As part of the new, focused, results-based approach, we have moved to a team-based performance compensation system focused on performance related to a single business outcome. The new team-based approach will go a long way toward eliminating the inherent conflict stemming from individual performance compensation based on multiple measures; we will promote and reward the performance of the team over the performance of individuals.

"We are all in the same boat", the CEO said. "And – the name of that boat is *Making Money and Securing the Future.*"

"One of the areas we have not discussed yet, is something we call 'flow'", said the intrepid, results-based consultant, adding to the list on the board. "Flow is a process measure, along with cycle time and productivity, both of which we have already discussed."

## Process Measures – Cycle Time, Productivity, and Flow

"Cycle time is about duration, while productivity deals with process capacity and process utilization, and flow is about the relationship between sales, starts, and closings. In the early days of even-flow production, we tended to look at these three measures as mechanisms of even-flow, particularly, the relationship between sales, starts, and closings.

"What we have since learned, is that even-flow – from an operating performance standpoint – is an operating outcome, not a driver." Then, she asked, "What were some of the problems we had with even-flow?"

"It was a tsunami", said the CFO.

"Let's say it was anything but even", said the VP of Construction. "We had a very hard time coordinating the rate of sales and starts with the rate of closings. In fact, we were much more focused on the rate of starts; in effect, we treated the slots in the Start Slot Matrix as sacrosanct."

"And – what happened?", she asked.

"The most glaring problem was the fluctuation we experienced in the level of work-in-process", he said. "At times, the variation from month-to-month, from quarter-to-quarter was incredible. And – you're right about something else, too – there was a corresponding level of variation in the durations of the Start-to-Completion process. Start-to-Closing process, also, if you expand the parameters of the process."

"We are going to talk more about even-flow later", she said. "We are also going to talk more about what are known as 'pull' systems.

"The concept of pull – of pulling work into a process at the rate of production on the downstream resource – is one of the mechanisms that drive even-flow. Even-flow is what we want. But, sometimes, we have to go against our intuition and instinct in order to obtain it. Actually, what we need to do is change our instinct and intuition. One of the lessons we are going to learn is that we cannot create balanced flow in a system that has balanced capacity."

"We have talked about production as a pipeline, and how we understand size, cost, and capacity", she said. "We have connected business outcomes to the operating performance measures that drive those outcomes. We have learned 'what happens to money' from an operational standpoint.

"There was more. There is more to come.

"I want to go back to something, back to the image of production as a pipeline. Remember this: There are two ways you can increase productivity without increasing designed production capacity. There are two ways you can increase the rate of Throughput, without re-sizing the pipeline."

The intrepid, results-based consultant moved to the erasable board, and wrote:

*Efficiency*

*Effectiveness*

"First – you can find ways to utilize the pipeline more efficiently, by taking out all of the errors, waste, delays, unwanted variation, and other non-value-added work. In essence, you can straighten and shorten the pipe. This is what your kaizen efforts have been about.

"Second – you can find ways to use the pipeline more effectively, by pulling the designed rate of Throughput through the system all the time, with a planned, finite, and controlled level of work-in-process."

The intrepid, results-based consultant addressed the team. "That's it for today. I want you to think about the new lessons we have learned", she said. "Find ways to connect the operating decisions you make every day to the business outcomes we want. Next time, we will pick up where we ended, and talk about how systems operate, how they behave, and how you manage them. You guys keep getting better and better. As always, you were a very attentive, very engaged group. And – I appreciate your efforts."

# CHAPTER III: SYSTEMS-THINKING

The intrepid, results-based consultant moved to the front of the conference room, smiled at her team, and said, "Good morning. So far, we have covered the 'picture' of production and how measures of operating performance are linked to the business outcomes we seek.

"Let's do a little review.

"What is the image we convey of a production system?"

"It is a pipeline", answered a superintendent. "The *size* of the production pipeline is determined by the number of homes under construction. The *capacity* of the pipeline is the rate of closings in relation to the planned, controlled, and finite number of houses under construction. The *cost* of the pipeline is what we pay every year, in operating costs and resources, to have the use of it. And – we incur the cost of the pipeline, whether we use it or not."

"Nice summary", said the Intrepid, results-based consultant. "Tell me, what is the relationship between the amount of work-in-process (WIP), cycle time, and output?"

"Little's Law tells us this: If you know two of these measures and you know the number of days in the period you want to measure, you can always calculate the third measure", replied one of the sales representatives. "So, cycle time is a function of work-in-process – the average number of houses under construction – divided by the number of closings for the period." The sales representative walked to the board, and wrote:

$$CT = 120 \text{ days} \quad WIP = 80 \text{ homes} \quad T = 240 \text{ homes}$$

$$WIP = (CT \times T) \div \text{Days}$$
$$(120 \times 240) \div 360 = 80 \text{ homes under construction}$$

"Work-in-process is cycle time multiplied by the number of closings. Output – or, what we call Throughput – is calculated by dividing work-in-process by average cycle time." She continued to write:

$$CT = (WIP \div T) \times \text{Days}$$
$$(80 \div 240) \times 360 = 120 \text{ days}$$

$$T = (WIP \div CT) \times \text{Days}$$
$$(80 \div 120) \times 360 = 240 \text{ closings}$$

"Just solve for 'X'.

"Also, the buffers related to these three measures – work-in-process, cycle time, and output – are the three ways a production system protects itself from variation and uncertainty. That's called the Law of Variability Buffering.

"And – yes – sales representatives have brains."

The intrepid, results-based consultant smiled, and said, "Tell me about size and growth, as it relates to RB Builders."

"Based on our understanding of capacity and size, RB Builders does not want to grow or become bigger", said the VP of Construction. "We don't want to increase production capacity. We want to better utilize our existing production capacity, by becoming more productive. It's about speed and velocity. It's about 'more-for-less', not 'more-for-more'."

"Excellent", she said. "Tell me about the importance and relevance of cost behavior in relation to changes in Revenue."

"Variable costing is central to our understanding of production capacity and productivity", offered CFO. "There are costs – the ones that are direct, truly-variable costs – that we want to *control*. There are other costs – the ones that are indirect, non-variable costs – that we want to *leverage*.

"The ability to leverage the costs and resources we will incur regardless of changes in the Revenue we generate is the key to utilizing production capacity and increasing productivity."

"Very good", she said. "Okay – talk to me about money. From an operational standpoint – what happens to money in RB Builders?"

"*Throughput* – basically, what we know as Gross Income – is all of the money we generate through closings", said another superintendent. "*Inventory* – basically our work-in-process – is all of the money we invest in what we intend to turn/transform into Throughput. *Operating Expense* is everything we spend to transform Inventory into Throughput. T-I-E.

"So do superintendents."

The intrepid, results-based consultant tossed the superintendent an erasable marker, and said, "Using only those three terms, what are the operating measures that drive business outcomes?"

The superintendent moved to the board, and wrote:

Cycle Time = Inventory ÷ Throughput
Productivity = Throughput ÷ Operating Expense
Inventory Turn = Throughput ÷ Inventory

"And, the business outcomes?", she asked.

"I know there are expanded ways to measure it, ways that remind us that economic return is a function of both margin and velocity. And – I know we have to deviate a little to make the terms work. But – not to get too confusing on this", he said, turning back to the board and writing:

Net Income = Throughput – Operating Expense
Return on Assets = Net Income ÷ Inventory

"Hoocoodanode?", murmured the sales representative, rolling her eyes.

"What we want is understanding", the CEO reminded everyone. "Every important measure of operating and financial performance can be calculated – but, more importantly, they can be understood – using three measures: Throughput, Inventory, and Operating Expense. T-I-E.

"If we focus on making daily operating decisions that maximize the rate of Throughput with a planned, finite, and controlled level of Inventory, and we do it by leveraging our investment in our production capacity, then we will have shorter, faster cycle times, higher productivity, and faster inventory turns. And – as a result – we will be more profitable, we will earn a higher economic return, and we will generate more cash."

"Well-said." The intrepid, results-based consultant continued. "Okay. So – how much work-in-process does RB Builders need?"

"We want the amount of work-in-process necessary (Necessary WIP) to produce our targeted closings within our targeted cycle time", replied the CFO. "Right now, we're close to the maximum level (Max WIP), which is determined by our Debt-to-Equity Ratio; it's not where we want to be."

"This idea of 'size' is deeply connected to the issues of capacity, productivity, speed, and velocity", said the intrepid, results-based consultant. "It is deeply connected to business outcomes – to higher profitability, stronger economic return, and higher cash generation."

"And – our intent is to continuously improve operating performance, so that our level of Necessary WIP becomes smaller and smaller", she added. "Okay, let's keep moving.

"Last time, we had a discussion about margin and velocity. We noted that Return on Invested Assets is a composite measure of economic return. It is an outcome of both the margin (Return on Sales) we earn, and velocity (Asset Turn) we generate. Again, margin and velocity work together to produce economic return. Margin is how much money we make on every house we build and close, and velocity is about how many houses we can build and close – of course, with a planned, finite, and controlled amount of work-in-process.

"However, while economic return (ROIA) is a composite of both measures, we noted that margin and velocity are driven by different aspects of the business, each with their own limitations or constraints. We tend to be focused on one or the other, depending on our job responsibilities, so we only appreciate the measure that is our job focus.

"So – other than what I have just said, how are margin and velocity different, and do we have to choose between them?"

"As you said, margin is about everything we do to make as much as we can on every house we build", said the VP of Sales, "while velocity is about everything we do to maximize the number of houses we can complete and close with a both a planned, controlled, and finite level of work-in-process, and a planned, controlled, and finite amount of resources to do the job.

"On the one hand, margin deals with marketing and selling, about having the right product, and extracting value from costs, which is why we want to control, and hopefully reduce, those costs. The limitation to higher margins is external to the production system; it is in the market. On the other hand, velocity is about production, about job scheduling, production management, and project management; it is about production capacity and productivity, about doing more for less – really, the things we have been learning in these sessions. The constraint to higher velocity is internal."

"Do we generally have to choose to prioritize our efforts on whether to increase Return on Sales and efforts to increase Asset Turn?", the intrepid, results-based consultant asked. "Do they adversely react to each other?"

"Not usually", answered the VP of Construction. "What can force the choice is circumstances where the limitation is pronounced, like what we expect next year. In 2008, we are going to need sales, and we are going to have more capacity than we can use.

"But – no – generally speaking, it's not a forced choice. We can generally exploit both. And, if there is a conflict, the referee is Throughput."

"The beauty of Throughput", said the intrepid, results-based consultant, "is that we get more of it as a result of both higher margins and higher velocity, and it does not distinguish between them. Throughput is Throughput. Gross Income is Gross Income. Contribution Margin is Contribution Margin. We can exploit both. Velocity produces its additional Throughput with fewer assets deployed, and we might have more control in overcoming internal constraints than we do in overcoming external constraints."

"Is that the only distinction?", asked the VP of Sales.

"There is one other distinction we should make", she replied. "I didn't mention it during the previous session, the session we are reviewing now.

"Let's start by agreeing that each dollar of Revenue that we 'spend' decreases Net Income. In that sense, a dollar of Revenue lost on lower margins is the same as a dollar of Revenue spent increasing velocity. But – everything you spend that doesn't create value reduces margin. It reduces margin related to a single house, and every house you build loses that dollar again. Over and over, on every house. In that sense, waste perpetuates itself.

"On the other hand, efforts to increase velocity – efforts to improve productivity – can be viewed as something of an investment, even if those dollars are expensed. Those dollars are spent once, and you reap the benefit of spending – or investing – those bucks on every house you are able to build now, that you couldn't build before.

"For both distinctions, velocity dollars are more valuable than margin dollars.

"Margin is residual Gross Income. That's why it's called Gross Margin. It is what remains after all variable costs. That is why you have to control and extract value from variable costs. If you don't extract value, variable costs reduce margins.

"And, tell me what you know about flow", the intrepid, results-based consultant continued.

"Flow is a process performance measure", said a sales representative. "The other two process performance measures are cycle time and productivity. Flow is about the relationship between periodic sales, starts, and closings. We want even-flow. But – even-flow is an outcome, not a mechanism.

"In the past, our rate of sales, starts, and closings have not been very even. One result has been a lot of variation in the level of work-in-process. You also said something about pulling starts into the system, and something about the difficulty of creating balanced flow in a production system that has balanced capacity."

"That's a good start", said the intrepid, results-based consultant. "We will talk a lot more about flow, pull systems, and balanced capacity. Last question. What are the two ways you can increase productivity without increasing designed production capacity?"

"The first way is to make the pipeline itself more efficient, by taking out all of the waste", replied the VP of Construction. "Straighten and shorten the pipe. The second is to use this more efficient pipeline more effectively, by maximizing Throughput and controlling the level of work-in-process."

"Very good review", she said. "We have covered a lot over the last two sessions, and you seem to have grasped it pretty well. Today, we are going to learn about what we call 'systems-thinking', about the universal, timeless, and self-evident set of principles that drive systemic behavior, including the behavior of the homebuilding production system we must manage."

On the erasable board, the intrepid, results-based consultant started a new list.

## Systems-Thinking

"In the first production principles and disciplines session we did, we said that we live in a world of systems. We do not need to become experts in general systems theory, but what do we mean?"

"For our purposes, I think we're talking about the relationship and interaction between dependent parts that comprise a whole", offered the CEO. "It is about the order of things, about relationships of cause-and-effect. It is about structure defined by parts and processes.

It is about understanding dependencies, and how things work together. You can say that systems are a generalization of reality, but that's the only way we understand that reality."

He wrote:

Natural order
Cause and Effect
Structure

"As long as I have been in homebuilding", he continued, "I have constantly been amazed at the way the industry tries to do things: Lurching from project to project, from deal to deal, acting like we've never done any of it before. Details without any context. Treating symptoms of the problem, not the problem itself. Trying to improve the performance of the individual parts, but not the performance of the whole."

"Sounds like whack-a-mole", said one of the sales representatives.

"When we use the term 'systems-thinking', what we mean is a way of reasoning, a way of problem-solving", said the intrepid, results-based consultant. "It is a way of thinking. Systems-thinking teaches us to think globally while acting locally. It teaches us to focus.

"The homebuilding industry, the housing and real estate market, and the local and national economies in which RB Builders operates – they are all part of a hierarchical system", she said. "The business environment in which we must operate – it's a system. These production principles and disciplines are part of that system.

"It is a hierarchy of goals, and the conditions required to attain them.

"RB Builders is not some loosely-connected set of independent and unrelated parts – a collection of processes, departments, systems, resources, policies, and other isolated pieces of a whole", she continued. "Like everything else, RB Builders is a *system* – a hierarchical set of interdependent parts that must work together to accomplish a stated purpose.

"Our understanding of production principles and disciplines is rooted in this way of systems-thinking", she explained. "Rooted in an understanding of cause-and-effect relationships, rooted in the interdependent nature of its parts, rooted in the ordered behavior of the business environment in which we must operate.

"It is rooted in the way things work. It is rooted in the way problems are solved, rooted in the way constraints are managed.

"Systems-thinking is about improving the performance of the entire system, not the performance of its pieces or parts – not any of the parts, not some of the parts, not even all of the parts, independent of one another."

The intrepid, results-based consultant made a numbered list on the flipchart, then paused and moved forward slightly for additional emphasis. "Listen. The way you improve a system is by focusing the improvement efforts on the two areas that determine the performance of the entire system:

1. Solving the core problem
2. Managing the constraint

"What is the problem? What is the constraint? Those are the two questions that have to be answered in order to improve the performance of any system. The system's core problem – its root problem – is the problem that causes the majority of the undesirable effects, the symptoms of the problem that permeate much of the system. Variation, for example.

"The system's constraint is whatever limits the system's overall capacity to do more of what it's designed to do, to produce more of what its purpose is to produce.

"Do you understand?"

The dazed looks and silence said it all.

Finally, the CEO stood up, walked to the erasable board, wrote a single word, and underlined it for emphasis.

Focus

The CEO paused before he spoke:

"What I think she wants us to understand is that this world of systems in which we must operate – this world of systems in which our understanding of production management must exist – does not provide us with promises of unlimited capacity, resources, capital, and opportunities.

"Which means that our efforts to improve, to solve problems, to manage constraints have to be prioritized.

"It means that our efforts to improve have to be focused. It means that solving some problems, eliminating some constraints, and exploiting some opportunities has to wait on solving more important problems, eliminating more important limitations, and exploiting better opportunities.

"It may appear that RB Builders has a lot of 'problems', and we can see all of the undesirable effects stemming from those problems", he said. "Likewise, it may appear that we have many so-called 'limitations'. However – not every problem and not every limitation we face is the one that determines the ability of RB Builders to achieve more of its purpose, nor is our ability to achieve our purpose determined by the sum of the capabilities and the capacities of individual teammates, departments, or any other parts of a whole.

"So what? This is what. Trying to improve most of the system will be counter-productive. It will consume too many resources and produce too little improvement."

"He's right", said the intrepid, results-based consultant. "Systems are *not* a set of equally-important-but-independent, related-but-isolated measures.

"The way we improve a system's performance – the way we improve production, improve operating performance, achieve desired business outcomes – is *not* by spending time and effort improving the outcome of each and every part. We cannot improve everything, everywhere, at the same time. Everything cannot be the root problem, everything cannot be the cause of the problem. Every limited or constrained resource cannot be the system's constraint.

"If we try to solve our problems this way, we will spend all of our time and effort treating the outward, visible symptoms of the problems, without ever resolving what caused it.

"If we try to manage constraints this way, we will spend all of our time and effort increasing the capacity of resources that do not limit the capacity of our production system, without ever identifying and effectively managing the constraint that does."

"She's exactly right", said the VP of Construction. "If we try to solve our problems and manage our constraints by dealing with things in isolation, we might improve performance in some areas and temporarily relieve a few of the symptoms and effects of the problems, but we will never solve the core problems, and we will never optimize the production capacity or maximize the productivity of the system. We will almost certainly add capacity, but it won't be capacity that makes a difference, capacity that increases Throughput."

"Well, the only thought I can add, is that I suppose it's better to be exactly right, than to be merely right", said the intrepid, results-based consultant. She continued, writing as she spoke.

## Systems = Chains

"Systems are a lot like chains", she said. "So – let's dig into that analogy. Would anyone care to explain the purpose of a chain?"

Clearing his voice, a superintendent stood. "The purpose of a chain – it would appear to me – is to sit in the back of my truck and rust", he said. "Not really. But – they do. Rust, I mean."

Looking at the expressions in the room, the superintendent became less amused with himself. "The purpose of a chain is to either hold something down, lift something up, or pull something somewhere."

"That pretty much covers it. And – how is the achievement of that purpose measured?", she asked. "How is the chain rated?"

"Chains should be rated according to their strength – according to their capacity to hold, lift, or pull something", he answered. "To my way of thinking, strength is what determines the value of a chain, which is it's ability to do work. A chain's value is it's ability to do what I bought it to do."

"Okay", she said. "If the value of a chain is its strength – its capacity to hold, lift, or pull something – then what determines that strength?"

The superintendent thought for a moment. "What determines the strength of a chain is whatever limits that strength", he said. "The strength of a chain is determined by the strength of its weakest link."

"So – where does the weight of the chain come into play?", she asked.

"The value of a chain has nothing to do with its weight, but weight has a lot to do with its cost", said the superintendent. "Just because a chain costs a lot, doesn't mean it has any value. To me, value is about the capability and capacity of the chain to do something, not how much it cost.

"I buy a chain because it's strong, not because it's heavy."

"If RB Builders was the equivalent of a single chain", the intrepid, results-based consultant continued, "it would never have more than one weakest link. As a homebuilding company, we might have a lot of what would be considered weak links, but we would probably only have one link that was materially weaker than all of the others.

"If a chain – it's purpose, strength, value, weight, cost, weakest link, etc – is a fair analogy of our production system, RB Builders would never have more than one constraint to better performance. But – our production system would always have to have at least one constraint; otherwise, it would have unlimited production capacity. In effect, we would have unlimited capability to accomplish whatever we wanted. RB Builders could produce as much as it wanted, whenever it wanted; and, we know that's not the case.

"In reality, RB Builders will probably have more than one constraint, because we have more than one chain of dependencies", she said. "In effect, it is more like dealing with 'systems within a system', with the enterprise lifted at different points, like at its four corners.

"In that case, we could say that it would have four chains, one attached to each corner and then to each other, not lifted from the middle by a single chain. No doubt, each of those four chains has one link that is weaker than any of its other links, but the system, the entire interdependent set of chains, will fail – will find the limit of its capability – according to the weakest of all of them.

"There are external, market-related constraints, there are internal, production-related constraints. There are constraints to higher margins, there are constraints to higher productivity. There are constraints on capital. But – all of these chains, with their own weak-and-weakest links, are part of the overall system. In some way, they are what determine our capacity and capability to improve performance – to improve what we do, how we do it, and what we accomplish.

"The particular chain that is RB Builders' production system", she continued, "is focused on generating as much Throughput as possible, within the parameters established by a planned, controlled, and finite level of capital, resources, and capacity. So – within the limitations imposed by a planned, controlled, and finite level of capital, resources, and capacity – RB Builders' production system is focused on whatever limits production.

"At any point in time, RB Builders will have very few internal, production-type constraints, very few things that really limit the rate at which we can generate additional Throughput – and, thereby, generate cash flow, make a profit, and produce a return on our owners' investment. And – one of those limitations will inevitably prove to be a bigger limitation than any of the others.

"That is where the focus must be, right now."

"I think the type of systems-thinking we have described here is going to be critical to everything we do, certainly to understanding production principles and disciplines", said the CEO. "It will give us focus."

"Like we said, systems-thinking is a way of thinking, a way of reasoning, a way of problem-solving", said the intrepid, results-based consultant. "It determines whether something that's 'not right' is really the problem. Systems-thinking distinguishes between a symptom of a problem and the real problem. It understands the difference between something that is the root cause of the problem, and something that is the effect of the problem."

As she spoke, she wrote:

Thinking
Reasoning
Problem-solving

"Systems-thinking focuses on managing constraints. Constraints are not bad, they are a fact of life for any system. As I said earlier, for a system to *not* have a constraint, it would have to have unlimited capacity. We know that's never the case; every system has capacity limitations.

"Constraints are the key to managing a system, to getting more out of it.

"To deal effectively with the capacity limitations imposed on the entire system by its constraint – by its bottleneck – the system needs to let the constraint set the pace and rhythm of production; it needs to let the constraint be the drum, let it be the pacemaker. That allows the system to exploit the constraint, by getting as much throughput out of it as it can.

"How does the system support the constraint?

"First, by subordinating the decision-making on every other resource to the needs of the constraint. Second, once the constraint is at capacity, the system elevates the constraint. It increases the capacity of the constraint, by switching it from some other non-constraint resource, or by acquiring additional resources; it finds more of it", she said. "It understands the nature of constraints, that some are physical, but the vast majority are simply limitations imposed by our internal policies and beliefs, imposed by our own way of looking at things.

"Systems-thinking digs beneath the surface, digs beneath the mere appearance of things, and constantly and repetitiously asks 'why?'. It does not settle for the first answer that it hears. It expresses legitimate reservation. It shatters erroneous assumptions and resolves conflicts."

Before concluding the discussion, she wrote:

*What is the core problem?*
*What is the solution?*
*How do we implement it?*

"From a production principles and disciplines standpoint, the recurring process – *the continuous process* – of improvement can be stated in very simple and straightforward terms: What are the core production problems? What is the simple, practical solution? How do we get to where we want to be from where we are now?

"With the review of the first two sessions, this third session took longer than I expected", she said. "Still – good stuff. This way of thinking and reasoning is very important. We will pick up from here next time, and talk about the production system.

"Thanks for hanging in there. Really good participation. I appreciate all of your hard work. Trust me – all of it will pay-off."

# CHAPTER IV:  THE PRODUCTION SYSTEM

The intrepid, results-based consultant burst into the conference room and strode to the front, talking as she went.  "I know.  I'm late", she said, between breathes.  "One of your owners has this annoying habit of calling just as I'm trying to get over here.

"Let's start with a quick review of what we have covered in the first two sessions, the visual image of the pipeline and the connection between operating performance and business outcomes.

"Short answers, please.  Anyone that knows the answer, just shout it out."  Amused with herself, she added,  "In *italics*.

"Size of the pipeline?"

*"The amount of work-in-process."*

"Capacity of the pipeline?"

*"The rate of Throughput that can be pulled through the pipe with a planned, controlled, finite level of work-in-process."*

"Cost of the pipeline?"

*"All of the indirect, non-variable operating costs, and resources associated with those costs, that we pay for the availability of the pipeline."*

"What is the mental model for increasing production capacity?"

*"'More-for-more'."*

"What is the contrasting mental model for improving productivity?"

*"'More-for-less'."*

"What mental model will you settle for?"

*"'More-for-the-same'."*

"Productivity is the relationship between _____ and _____."

*"Productivity is the relationship between what a system/process produces and what it consumes in the process."*

"Assets are _____ and expenses are _____."

*"Assets are transformed or converted, Expenses are consumed."*

"Little's Law and the Law of Variability Buffering – what do they say?"

*"Little's Law describes the relationship between cycle time, work-in-process, and output, and tells you how to calculate those measures. The Law of Variability Buffering describes the three ways a system will protect itself from variation, through a combination of higher inventory, longer duration, or excess/unused capacity."*

"What is RB Builders' goal?"

*"To make money now – and in the future."*

"Is that goal the same as the company's purpose, its passion, or its dream?"

*"No, but it's a strong necessary condition."*

"Okay. So – is 'making money' the way RB Builders measures success?"

*"Partly. There's more to the way we measure success than 'making money'."*

"Good", she said. "From an operational standpoint – what happens to money?"

*"Only three things: We generate it. We invest it. We spend it. Throughput, Inventory, Operating Expense. T-I-E."*

"Give me the three measures of operating performance you can calculate, using only the terms T-I-E."

*"Cycle Time, Productivity, and Inventory Turn."*

"What are the formulas?"

*"Cycle Time is Inventory divided by Throughput (multiplied by the number of days in the period); Productivity is Throughput divided by Operating Expense; Inventory Turn is Throughput divided by Inventory."*

"The reciprocal of Cycle Time is _____ _____."

*"Inventory Turn."*

"Okay. What financial measure on the Income Statement most closely relates to Throughput?"

*"Gross Income."*

*"I thought it was Contribution Margin."*

"You're both correct.  What are the two financial measures you can calculate with T-I-E?"

*"Net Income and Return on Assets."*

"Formula?"

*"The formula for Net Income is Throughput minus Operating Expense.  The shorter formula for Return on Assets is Throughput minus Operating Expense, divided by Inventory.  A more insightful version is the DuPont formula, which is a composite measure that breaks Return on Assets into its components."*

"Return on Assets is a composite of which measures?"

*"It's a composite of margin (Return on Sales) and velocity (Asset Turn)."*

"Please, explain.  Explain the relationship."

*"Margin is how much we make on every house we build.  Velocity is the rate at which we can build houses with a planned, finite, and controlled level of work-in-process and production capacity.  They focus on different aspects of the business."*

"Which aspects?"

*"Margin is marketing-related.  Velocity is production-related."*

"Where does margin focus?"

*"Margin focuses on product development, upstream and downstream marketing, sales.  It is a reflection of pricing and job costs, which involves partnering.  It is a function of plans, elevations, options, and specs."*

"Where does velocity focus?"

*"Velocity focuses on job scheduling, production planning, and project management.  It's about doing 'more-with-less'."*

"Where is the limitation to higher margins and higher velocity?"

*"Generally, the limitation to higher margin is outside the production system – it's in the market.  But, the constraint to higher velocity is usually inside the production system."*

"Is the focus on either margin or velocity a choice?"

*"Everything is a choice, and we are responsible for the outcome of our choices.  But – the answer is no.  Except under dire or extraordinary circumstances, we can focus on both margin and velocity."*

"How do we want to manage our direct, truly-variable costs?"

*"We want to either further reduce them, or we want to extract more value from them. Better still – we want to do both. We want to further reduce them and extract more value from them."*

"And – how do we want to manage indirect, non-variable costs?"

*"Since we plan to spend the money anyway, we want to better leverage those costs. We want to use them as fully as possible. We want to produce as much as we can with them."*

"How do we measure the performance of a process?"

*"We measure cycle time, productivity, and flow."*

"Flow? What kind of flow?"

*"We want an even rate of closings, which creates an even rate of starts, which is supported with a sufficient rate of sales. But, even-flow is an outcome, not a mechanism."*

"Outcome-not-a-mechanism? What does that mean? Please, explain."

*"It means that we can't achieve even-flow with even-flow. We can't mandate even-flow; we have to achieve the outcome of even-flow. We have to pull starts into the system, and we need to unbalance the capacity of the system, so that we manage variation and pace the system."*

"Interesting", said the intrepid, results-based consultant. "Tell me about the amount of work-in-process we need, and tell me about size and growth."

*"We want Necessary WIP, trending towards Minimum WIP. We want to maintain the amount of work-in-process necessary to support our targeted rate of closings at our targeted cycle time.*

*"Since size relates to both work-in-process and capacity, we don't want to grow at all. We want to become a faster, more productive homebuilding company, not a bigger homebuilding company; we want to increase productivity, not increase production capacity.*

*"We want our operation to be about 'more-for-less', not 'more-for-more'."*

"That was outstanding", she said.

"In this session, we are going to continue learning about systems. Last session, we learned about systems-thinking, about the ordered reasoning and the focus that enables us to solve core problems and manage constraints. So – let's quickly have a review on that knowledge, as well.

"Speak *italic*, please.

"We say that systems-thinking is a way of _____ - _____, _____, and _____. Three things, and the first one is hyphenated."

*"It's a way of problem-solving, thinking, and focusing."*

"In what relationships and approaches is systems-thinking rooted?"

*"Systems-thinking is rooted in cause-and-effect relationships, in parts that have interdependent relationships. It is rooted in ordered behavior, in the manner in which things work. It is rooted in the way problems are solved, and in the way constraints are managed."*

"How do you improve the performance of a system?"

*"Solve the core problem, the root problem. Manage the system constraint."*

"The analogy of systems as chains. What determines a chain's value? What determines its cost?"

*"Its value is its strength. Its cost is its weight; weight adds no value."*

"What determines the strength of the chain, and therefore, its value?"

*"The strength of the chain is determined – meaning, limited – by the strength of the chain's weakest link. The chain could have plenty of weak links, but it will fail at its weakest link, and there is only one."*

"Constraints can be located _____ and _____."

*"Externally and internally; though, not usually in both places at the same time."*

"How does the system support the constraint?

*"First, it subordinates everything to the needs of the constraint. Second, if the constraint is at capacity, the system elevates it, by increasing its capacity or finding more of it."*

"In terms of a production system, describe the process of continuous improvement. Three steps, please."

*"One, figure out what the core production problem is. Two, find the simple, practical – the elegant – solution. Three, determine how we get to the elegant solution from where we are right now."*

"You people are nothing short of brilliant."

"Today, we are going to learn how to apply systems-thinking in the context of a production homebuilding system. We are going to see how the production system is put together, and how it is managed. We are also going to talk about the 'things we manage as a system', and what I mean by that is, how we manage processes, value streams, and project portfolios.

"Let's begin with the definition of a system. Anyone?" Glancing toward the CEO, she said, "I believe you mentioned that you have recently started to think in this area."

"Yes – I have", replied the CEO.

He walked to the erasable board, selected a marker, and started a list, beginning with the following statement:

*System = a set of interdependent parts working to accomplish a purpose*

"The definition of a system goes back to what we discussed in the previous session", he said. "A system is a set of interdependent parts that must work together to accomplish a stated purpose. The system – this 'set of interdependent parts' – could be a process, could be a value stream, could be – as you said – a portfolio of projects."

"Tell me about the differences – and the similarities – between a production system, a process, and a project portfolio", said the intrepid, results-based consultant. "Do you see them as being the same?"

"No, I don't see them as being the same, at all", he replied. "As you said, there are both differences and similarities.

"I see a production system as a repository for a number of processes, value streams, and project portfolios. The production system provides the understanding and tools – the mental models, standards, routine problem-solving, etc. – that allow us to manage production."

Turning back to the erasable board, the CEO continued his list:

*Production System = repository of processes and projects*

"The production system is the underlying thinking – the understanding – and it is the repository for everything else, which are essentially different classes of workflow . . ."

" . . . and since the workflow horse has been let out of the barn," said the intrepid, results-based consultant, taking over the discussion from the CEO, "I need to provide some context.

"The most basic, most universal proposition of business is simply this: The reason for an enterprise's existence – the way it makes money – is through the value that it delivers to customers and other stakeholders. That value is only delivered by the work that the enterprise performs, and that work has to be performed in some method of workflow.

"Those methods of workflow exist, whether enterprises – RB Builders or anyone else – are intentional about them or not. Those methods are descriptively termed process management, project management, and case management."

She started a separate list.

"There was once the prevailing view that workflow was all about processes – that workflow was synonymous with process. The sequence of the proposition underlying business process improvement was commonly stated as value-work-process. But, the spectrum of workflow is broad, and a single approach to performing value-adding work does not fit every enterprise situation, or every industry space."

*Project Management = longer schedules, lots of dependencies, different WBS*

"In construction, workflow is project management with embedded, repeatable processes. All contractors, including homebuilding companies, are Project Management Organizations; moreover, because they manage multiple projects, they are actually Project Portfolio Management Organizations.

"Compared with the other methods of workflow, project management has longer timeframes with a lot of task dependency and resource contention; although individual projects – i.e., jobs – have similarities, their work breakdown structures can vary significantly, depending on the project requirements. So, they tend to be treated as individual projects managed in a portfolio.

"PMOs schedule projects, not the process."

*Process Management = continuous, repetitive, definable, standardized*

"On the other hand, the operations of most manufacturing companies are about process management – continuous flow, sometimes single-piece, very repetitive, very definable, very standardized, often, very transactional.

"There is a distinction between cross-functional process workflow and workflow performed in sequence by one person or department, which is a set of procedures. There is also a distinction between processes, which provide an in-depth, largely internal understanding of workflow, and value streams, which provide an external, higher-level view of workflow, resource capacity, material flow, and information flow.

"Having drawn those distinctions – processes, procedures, and value streams are of the same genre; project management is a different animal, and process-centered enterprises have workflow management requirements that are distinct from those of PMOs. If you place process management and project management on a spectrum, they tend to be at opposite ends."

*Case Management = unstructured, ad hoc, collaborative, doc intense*

40

"Then, there is the emerging discipline of case management, which is not on the single plane/spectrum with processes and project management. In a sense, process, case, and project management triangulate. They are distinct. Case management has been around a long time in some industries, like healthcare, insurance, and legal; increasingly, case management is what you see in mortgage banking.

"In situations that have an unstructured progression of workflow, that are more ad-hoc – that is, less pre-defined – case management is becoming an alternative to a pure process approach. Case management also applies to workflow situations that are document-intense, that need to share documents in the same digital folder, that require real-time collaboration, and that involve physically separated, remote, and independent resources."

The intrepid, results-based consultant glanced around the room, stopping at the CEO. "Relax. It's just context. It will help all of this make more sense. Trust me.

"Pardon the interruption."

"I look at processes, value streams, and projects as the end-to-end sets of steps required to turn inputs into outputs that meet specifications", the CEO continued. "As you said, processes tend to be continuous. They are tightly tied to standards. If a process is a very short set of steps performed by just one person, I agree, it is best to just call it a set of procedures, and the term 'process' should be reserved for longer, more complex, cross-functional sets of activities. In particular, we try to make procedures visual.

"Processes can be transformational, but, usually, they are, as you say, transactional."

The CEO continued his list:

*Production System = repository of processes and projects*
*Processes = end-to-end steps, input into output, internal perspective*

*Procedures = short, simple, one person, no handoffs*
*Processes = longer, more complex, cross-functional*

"Might not be the best descriptions, but they will do. We've done some work in the process area. We have been documenting our processes as hierarchical process models using a preferred method of process notation.

"Then, as you said, we have value streams", he said, adding to his list.

*Value Streams = higher-level, external perspective, info, data, kaizen*

"Value streams are a much higher-level view of a process. Value streams are directly associated with Lean Thinking, which describes the production method that grew out of the Toyota Production System. Value streams present a picture of supply and demand, which is distinctive. The other distinguishing characteristic is the inclusion of suppliers and subcontractors, which results, as you indicated, in value streams having an external perspective, whereas a process provides a purely internal perspective.

"Enterprises that extensively use them document their value streams in what is called a value stream map. A value stream map shows information and material flow, in addition to workflow activities, and it attaches a lot of data to the activities, which adds considerable insight.

"Coming out of the TPS and Lean, Value Stream Mapping – VSM – separates improvement efforts into flow kaizen, which is about improving the flow of the entire value stream and is mostly a management issue, and process kaizen, which is about eliminating waste in the process and involves more the individual or small team. We would probably concur with that distinction, but we also have a varying level of sophistication on this, so, sometimes kaizen is just kaizen as far as we're concerned."

The CEO completed the list.

*Project Portfolio = Managing projects in a multi-project environment*

"Then – there are project portfolios", he said. "As you said, there is the issue of how you manage projects in a multi-project environment, which is what homebuilding production really is. That's an important distinction. Without going into details, homebuilding is different than manufacturing.

"We look at our overall value delivery process as the prospect-to-closing process", the CEO said. "This process has a lot to do with flow, cycle time, work-in-process, capacity – everything we have already talked about. Within the prospect-to-closing process, there is an end-to-end sequence of sub-processes known as prospect-to-contract, contract-to-start, start-to-completion, and completion-to-closing.

"That's P-T-C, C-T-S, S-T-C, and C-T-C, for short."

He wrote it on the erasable board.

*Prospect-to-Closing = P-T-C – C-T-S – S-T-C – C-T-C*

"All of these sub-processes display characteristics we associate with cross-functional processes. Except one. Except the sub-process in which we actually produce the product. That process is the start-to-completion process. Start-to-Completion is a process, but it behaves like a portfolio of projects, and it has to be managed in that manner.

"When we are managing pre-start and post-completion activities, contracts, closings, etc., it's okay to treat that workflow as a process; however, not Start-to-Completion.

"Essentially, S-T-C is a portfolio of projects moving through a process, a process that is surrounded by other processes, each of which is a sub-process of the primary value delivery process. And – somewhere outside of all that are value streams. That is a comprehensive view.

"But, the only aspects of the start-to-completion process that actually behave like a process are the milestone events, like walkthroughs or inspections, and the recurring tasks, such as moving the job schedule or authorizing variance purchase orders. I'll show you what I mean."

The CEO returned to the board, and made another list.

Critical Path
Logistics
Projects, Processes, Value Streams: What do we schedule?

"I could go into a lot more detail, but the single-piece or continuous flow of most processes have nothing like the critical path of activities that projects have", he said. "You also have to think about the sheer logistics of how we build homes, the environment in which we build homes, and who does the actual work. And, resolving resource contention, in which the same resource is required on different jobs at the same time, is much more a part of the project portfolio environment than it is the process environment.

"Finally – look at what we schedule. As you said, we schedule jobs. Not the process. Not the value stream. Jobs are projects. Our work-in-process is a network of jobs/projects.

"As you said, value is created by work, which is, in turn, performed in some method of workflow – processes, projects, or cases. And – if you look at processes as . . . "

"Are you about finished?", asked the intrepid, results-based consultant.

"No, I'm not", said the CEO.

"Not to create more confusion, but the term 'process' applies to more than the workflow performed in processes; the term is also associated with the methodology used to improve something, as in 'a process of continuous improvement', as in 'business process improvement'. It would seem that every discipline – TQM, Lean, Six Sigma, TOC – has an improvement methodology characterized as a process. The use of the term as a methodology is a good example of a transformational process, as would be something like the process for evolving subcontractors and suppliers into building partners and market partners.

"So – when you hear the term 'process', you just have to consider how it's being used. It is obviously used in connection with workflow, but there are methods other than processes for performing work.

"Sometimes, it doesn't have anything to do with workflow."

"The best production models – the best-known, most highly-regarded ones – all come from manufacturing", said the CEO. "However, homebuilding is a different kind of management situation. It has different parameters. Plus – production principles and disciplines are not as highly-developed in homebuilding as they are in other industries, due, in part, I suppose, to our finely-honed deal-driven mentality."

"Meaning what?", asked the intrepid, results-based consultant.

"I know what he means", said a superintendent. "He means land deals that turn into community developments. It's really finely-honed lurching. But, the deal-driven mentality isn't restricted to so-called deals. Literally, it's everything we do. RB Builders has only recently gotten true process religion, and the still, small voice of deep systems spirituality has only been whispering to us for what? About a month?

"We tend to do everything like it is the first time, like we have never done it before", he said. "We need to become more process-centric."

"He's right", said the CEO. "That is slowly changing. But – back to what I was saying. Since we have to manage in our world, the world of homebuilding, we know we need to find a way to adapt the best aspects from these models to the specific set of requirements and circumstances we face. We need to be able to manage processes, manage appropriate aspects of value streams, as well as manage projects, as the situation dictates."

"Accurate summary", said the intrepid, results-based consultant. "Homebuilding is a different scenario than most manufacturers or service providers face. It has a different set of requirements. However – while processes, value streams, and project portfolios are all systems – projects are different from either processes or value streams. Project management has more in common with process management than process management has in common with project management. If that makes any sense."

"No", said the CFO. "It doesn't make any sense. What do you mean?"

"Well", she answered. "It's just that there are more elements about processes that underlay our understanding of projects, than vice-versa. Processes are more foundational. It's kind of like first-level learning. It is easier to approach project management from an understanding of processes, than it is come at process management from an understanding of projects.

"So – this is what I'm getting at. Let's set project management aside, for now.

"Value stream maps can help us 'see' improvement opportunities more clearly, in both processes and project portfolios. But – for a more indeterminate period – let's also set aside value streams.

"For right now – let's talk about production management from a process perspective.

# CHAPTER V:  PRODUCTION PROCESSES

The intrepid, results-based consultant glanced around the room to make sure everyone had arrived and was seated.  "Let's pick up where we left off", she said.  "From a production perspective, what do we need to understand about processes?"

"I've been keeping a list of things I think we need to understand and do", said the VP of Construction.  "The first area on my list is variation and uncertainty.

"I know we have embraced, to a significant degree, the principles of Lean, and one of those principles is to remove waste from the value stream.  I know that variation is a form of that waste, and Lean attacks waste with kaizen and PDCA.  I know that, in some improvement circles, the principles of Lean and Six Sigma have been blended into Lean Six Sigma, and that the Six Sigma portion of LSS focuses on removing variation from processes.

"But – if this is process religion, then the demon is named Murphy.

"If something can go wrong, it invariably will go wrong.  And – it will go wrong at the worst possible time.  We see it over and over and over again, and the result is uneven flow and long cycle times.  Maybe some other stuff I don't even understand yet."

"I take it that you think Murphy is somehow different, or is deserving of special consideration?", the intrepid, results-based consultant asked, writing as she spoke:

*Variation/Uncertainty*

"Yes and no", he said.  "There are two kinds of variation.  There is common cause variation and there is special cause variation.  The attributable variation – or common cause variation – is largely what Lean attacks with its kaizen and PDCA.  The natural variation – or special cause variation – is different;  it is more akin to uncertainty, and usually it is beyond our control.  I want to know how our system protects itself – how it buffers itself – from both variation and uncertainty.  And – since we are including natural occurrence – we should look at the behavioral aspect of human nature."

"Yes, having an understanding of variation and uncertainty is important", said the intrepid, results-based consultant.  "Variation lies at the heart of every other production principle, because it affects everything.  The elements of a production system – the elements of production processes – are largely driven by the need to control variation and deal with uncertainty.

"Incidentally, there is a difference between variation and uncertainty.  The common cause type of variation is a reflection of our ability – or our inability – to achieve repeatable results, while uncertainty, what we call special cause variation, is a reflection of not knowing what the results will be.

"Uncertainty is about risk. It is unexpected. Uncertainty is what causes results to extend beyond the range of normal variation."

Looking at the dazed expressions on the faces of the group of superintendents at the end of the conference room table, she shook her head in disbelief.

"The difference between variation and uncertainty is something like this", she explained. "When you go to Mike and Ed's for lunch, sometimes it takes you 45 minutes, sometimes it takes you an hour. That is common cause variation, and you won't change it, unless you change your routine – unless you change the process. However, sometimes lunch at Mike and Ed's can take 90 minutes – traffic, iced tea dispenser is empty. That is special cause variation, or uncertainty. You can't predict it.

"Natural human behavior is actually learned or acquired, but it is such a common, prevalent response, it might as well be termed natural. It is significant", she said, looking toward the VP of Construction. "You brought it up. Do you care to elaborate?"

"Sure", said the VP of Construction. He walked to the erasable board, cordoned off some space, and started his own list:

Padded Durations

"Here's what I mean", he said. "People begin, by padding the estimated duration of tasks, in order to provide more certainty and higher reliability. Let's use our framing contractor as an example. In his mind, it goes something like this: 'If everything goes right, it's only going to take me 10 days to frame this house, but, just to be on the safe side, I'll tell RB Builders it will take me 20 days'. This padded duration – this hidden safety – is unspoken.

"It's not intended to be deceptive or anything, but it's still hidden. But – then – the other tendencies of human behavior conspire to consume all of that hidden safety, and it winds up taking 20 days or longer to frame the house. If you add up all the hidden safety in every task, across an entire job schedule, it's no wonder we are looking at 120 day job schedules.

"Of course, we don't complete houses in 120 days. We complete them in closer to 180 days, even though we have this gut-instinct that we should be able to build every house in less than 90 days."

"What 'other tendencies' are you talking about?", she asked.

The VP of Construction added to his list:

Padded Durations
Student Syndrome
Multi-tasking
Parkinson's Law
Murphy

"The framing contractor in this case doesn't really start the job, until he's used up most of his safety. I mean, he may start it, but he doesn't do much work. It's the same thing students do with their homework assignments and test preparation – they wait until the last minute to start.

"In the meantime, he looks at his schedule, sees the extra days, and multi-tasks the jobs, meaning that he will spread each of his framing crews between more than one job, so they will stay busy, rather than just adhere to a 'start one – complete one – start one' routine.

"Add to that, a general tendency for every task to expand to the time that it is given. It's known as Parkinson's Law. It's not physics, it's really just a tendency, a behavior. Anyway – even if we mandate an early start – the duration of the task tends to expand to whatever completion date it's given. The delays accumulate, but the early completions are lost.

"As soon as 'we' do all of this – consume the hidden safety, multi-task, expand to the allowed time – then Murphy shows up, and the job finishes late. Or, that particular job finishes on time, because we expedited it, but that action then negatively affects every other job in the system."


"That was sort of impressive", said the intrepid, results-based consultant. "What you have so accurately described does affect processes – it affects anything involving tasks, schedules, and resources. But – it really affects the way we manage projects. We will talk more about it within the context of project management.

"What else is on your list?"

"This one could be related to variation", said the VP of Construction. "We need to know more about how this pull scheduling system works, the one you mentioned during the session on connecting business outcomes to operating measures.

"And, we need to understand more about this paradox you mentioned – about how we have to learn to unbalance the capacity of a production system in order to create balanced flow – that is, to create even-flow."

"Anything else?", she asked.

"I have one", said a superintendent. "I don't know if this is a process issue or a project management issue, but these production principles and disciplines need to translate into shorter, more reliable job schedules. Otherwise, it simply won't be worth the effort. I take it that we shouldn't just be scheduling individual jobs, but the entire system, as well. That's difficult, with each superintendent scheduling his own jobs, and competing for resources. My guess is, all of this stuff is somehow connected, so you can just add mine to the list."

She continued her list of the team's questions regarding production processes.

*Variation/Uncertainty*

*Pull Systems*

Unbalance Capacity to Balance Flow
Scheduling

"Does that capture it?"  The intrepid, results-based consultant thought about the best way to help RB Builders understand how production systems are put together and then managed. "We have covered a lot of material on production principles and disciplines, and there are a number of ways to explain how production systems work", she finally said.

"The best approach would be to cast it as two elements."

Turning to the erasable board, she wrote:

1.  Variation/Uncertainty
2.  Scheduling

"The first element is how a production system manages variation and uncertainty, which is the same for a process, a value stream, or a project network.  The second element is how a production system is scheduled, which is similar for processes and value streams, but different for project networks.  These two elements play off one another, with the second element – the scheduling element – being the key to managing the first element, i.e., the element of variation/uncertainty.

"Everything else related to production management fits under these two core elements," she said, looking toward the CEO.  "In fact, when you characterized RB Builders' production system earlier as a repository for certain processes, it is one of those processes – the manner in which RB Builders plans and manages production at the community level – that is most closely associated with the elements of variation, uncertainty, and scheduling."

"I have a question", said the VP of Sales.  "In one of the previous sessions – the one on systems-thinking – you said that an understanding of systemic behavior would allow us to solve problems and manage constraints.  Correct me if I'm wrong, but we are now getting ready to talk about scheduling, which involves resources, which gets into the subject of constraints.

"But – when will we talk about solving core problems?"

The intrepid, results-based consultant rummaged through the flipchart until she found the list to which the VP of Sales was referring, and made a change.

1.  ~~Solving the core problem~~  later
2.  Managing the constraint

"You're right", she said.  "Both problem-solving and managing constraints are part of how we manage a production system.  But – we will talk about problem-solving later, outside of production principles and disciplines, because it is a much broader subject.

"Solving problems is an important part of how we manage every aspect of a system, including how we solve production-related problems. Someday, we will get into the full range of root-cause problem-solving methodology, from the simple techniques associated with Plan-Do-Check-Adjust (PDCA), to the methods for resolving very complex problems.

"But, not right now.

"In these sessions, we are going to focus on the other area that determines the performance of the entire system. We are just going to talk about how we better manage whatever it is that is preventing the system from producing more with what it has. That is a discussion of constraints.

"Let's talk about variation and uncertainty", said the intrepid, results-based consultant, underlining that point on the erasable board.

1. *Variation/Uncertainty*
2. *Scheduling*

"It was mentioned earlier that variation is a form of waste, because, to the extent that it doesn't add value, variation is wasteful. That thinking largely comes from the Toyota Production System and the Lean Production methodology that TPS later spawned. But – Taiichi Ohno did not include variation as a form of *muda*, the term we use for waste. Instead, Ohno opted to use the term *mura*, which means unevenness, and to associate that term with the separate principle of stability; in Taiichi Ohno's view, production processes need stability, and variation causes the opposite, it causes instability.

"Interesting, too, in both Lean Production and the TPS, the principle of stability sits between the principle of waste and the principle of standardization, which includes the concept of standards, visual management, and problem-consciousness, which are linked to PDCA problem-solving.

"So – there is that distinction that would seem to differentiate variation from other forms of waste", she said.

"I think there is a reason Mr. Ohno didn't place waste and variation in the same category. There is no possibility of having either zero waste or zero variation; the goal is to reduce waste and reduce variation, which is achievable. However, given their inherent characteristics, a desire to eliminate waste is more reasonable than a desire to eliminate variation."

Turning to the board, the intrepid, results-based consultant wrote:

*Distinctions – Waste v. Variation*
*Production Physics: Law of Variability Buffering*

"Then – beyond that distinction – there is the series of basic laws of production physics that directly or indirectly form our understanding of variation. Five laws, to be exact, but two in particular", she said. "The first law, called the Law of Variability, states that increasing levels of variation progressively degrade the performance of the production system.

"The second law, the Law of Variability Buffering, is one we previously discussed. It says that variation will always be buffered by some combination of inventory, capacity utilization, or time – will always be buffered through a combination of higher work-in-process, excess/unused capacity, or longer durations."

The intrepid, results-based consultant added new statements to the list on the erasable board:

Buffers = Protection
Three ways:
1. Higher WIP
2. Excess/Unused Capacity
3. Longer Duration

"So", she continued, "If all RB Builders does is attack variation by attacking waste – in the form of errors, rework, etc. – and it fails to directly attack the variation that causes instability, then it will have to live with a production system that protects/buffers itself with some combination of additional work-in-process, longer-than-necessary cycle times, or wasted capacity. RB Builders' production system will default to longer durations. Time is the self-determining buffer, the 'buffer of last resort', so to speak. So, if the company does nothing about variation, yet it limits work-in-process and capacity, the inevitable result will be long cycle times. It's guaranteed.

"Buffers – high levels of work-in-process, long cycle times, and excess/unused capacity – allow the system to compensate for variation, however, regardless of the combination in which they occur, they all result in lost Throughput.

"And – the true cost of variation is the financial throughput – the Gross Income – RB Builders surrenders to that variation.

"You begin to get a sense that variability is a very big deal", the intrepid, results-based consultant continued. "Jack Welch, the former CEO of General Electric – which had picked up the Six Sigma mantle from Motorola – use to say, 'Variation is evil'. Some variation and uncertainty is natural, and – in some cases – it is necessary and planned. But, the instability that variation and uncertainty cause is a decidedly evil form of waste.

"In process religion, the gods of production will not be mocked", she said. "If all that RB Builders does is attack variation by attacking waste, in the form of errors, rework, etc., and the company fails to also directly attack the variation that causes instability, then – I'll say it again – it will have to live with a production system that protects itself with some combination of too much work-in-process, longer-than-desired cycle times, or excess/unused capacity.

"And, the result will be reduced throughput."

"But – isn't protection a good thing?", deadpanned a superintendent. "After all, shouldn't we be practicing safe production?"

"Yes, protection would be a good idea", answered the intrepid, results-based consultant, equally deadpan. "Especially for boys like you, who should be worried about contracting a PTD.

"Put it this way: Which one of these three Productionally-Transmitted Diseases would you like to have? Which combination of longer-than-necessary cycle times, higher-than-necessary levels of work-in-process, and excess and unused capacity resulting in lower-than-possible rates of throughput would you prefer to contract?

"I hear they are all really painful. So – you might want to consider abstaining from variation promiscuity.

"Some level of variation and uncertainty is natural, inevitable, and unavoidable", she continued, satisfied with the effect. "We have to buffer that. In addition, some variation is necessary, just to protect ourselves in the marketplace. For example, we don't offer only one floorplan and elevation. Lastly – whenever we are protecting the output of the system from variation and uncertainty – some level of protective capacity or buffering is needed.

"However – protecting a system from variation comes at a cost, and to the extent that the variation that necessitates the buffering is unnecessary, avoidable, excessive, or uncontrollable, it's a bad thing."

"You're right, this does sound like physics", said one of the superintendents. "It's nice in theory, mesmerizing stuff, but give us something we can use."

"Let's see if an example can help", said the intrepid, results-based consultant. "In a previous session, someone said RB Builders closed 200 homes in 2007, on an average work-in-process of 100 houses, which is also the baseline for 2008. But, everyone also agreed that RB Builders' production system should be capable of producing 240 closings on 100 units of work-in-process. Later, someone else mentioned that the building schedules averaged 120 days."

Moving to the erasable board, she constructed the following data table:

| Year | WIP | Closings | Cycle Time |
|------|-----|----------|------------|
| 2005 | 100 | 225 | ? |
| 2007 | 100 | 200 | ? |
| 2008 | 100 | 240 | ? |

"Someone calculate RB Builders' cycle time", she said. "How many days?"

"For which process?", asked the CFO.

"S-T-C", replied the intrepid, results-Based consultant. "Start-to-Completion."

One of the sales representatives looked up from her calculator, and said, "Okay, if I am correctly using Little's Law to calculate cycle time, in 2005, our cycle time was 160 days. In 2007, it was 180 days. And, for 2008, we are targeting 150 days."

The intrepid, results-based consultant completed the cycle time column with the calculated cycle times, but added two more rows.

| Year | WIP | Closings | Cycle Time |
|------|-----|----------|------------|
| 2005 | 100 | 225 | 160 |
| 2007 | 100 | 200 | 180 |
| 2008 | 100 | 240 | 150 |
| | ? | ? | 120 |
| | ? | ? | 120 |

"Now", she said. "Tell me what RB Builders' production system looks like with a cycle time of 120 days."

"There are two ways to look at it", said one superintendent. "We could be closing 240 homes with 80 units of work-in-process. Or – we could be closing 300 homes with 100 units of work-in-process."

The intrepid, results-based consultant added the new calculations.

| Year | WIP | Closings | Cycle Time |
|------|-----|----------|------------|
| 2005 | 100 | 225 | 160 |
| 2007 | 100 | 200 | 180 |
| 2008 | 100 | 240 | 150 |
| | 80 | 240 | 120 |
| | 100 | 300 | 120 |

"Remember our discussions on margin and velocity?", she asked. "This is where that thinking comes into play. There are two ways that RB Builders can increase the amount of Throughput – the amount of Gross Income – it generates. Margin is how much money we make on every home we close, and velocity is about how many houses we build and close."

On the erasable board, she wrote:

Margin
Velocity

"Most of the time, there are improvement opportunities that permit us to attack margin and velocity simultaneously", she said. "Like the DuPont formula used for calculating ROA, Gross Income is a composite of both margin *and* velocity. The aim is to have the most effective blend of both. As objectives, margin and velocity don't often conflict, although there are times when we might be better served focusing more on one than the other.

"Under some circumstances, we can be forced to make a choice on where to focus our effort, usually when we are facing an external, market-based constraint. Why one and not the other? Why typically an external constraint, and not one that is internal? Contrast where we are now, at the end of 2007, to the period we have recently emerged from. Where was the constraint in 2004-2005? It was in our production system. It was internal; RB Builders' internal capacity couldn't meet the demand of the market.

"The velocity part of the choice decision lies in how well RB Builders is utilizing its true production capacity", said the intrepid, results-based consultant. "The margin part is determined by the condition of the housing market, and whether that market is going to allow us to use – to economically leverage – that capacity. We can control truly-variable direct costs, and we can extract more value, but – at the end of the day – the market gives, and the market takes away.

"And – a lot of people are thinking 2008 might be that kind of year, barely a year removed from the recent, final, halcyon days of the Age of Homebuilder Entitlement. The prospective irony in this situation is that diminished demand – a market/external constraint – can later result in reduced capacity, if, for example, construction labor leaves the workforce, which would exacerbate the internal constraint in any recovery; as you can see, it is connected.

"The point is, we have to learn to manage the relationship between margin and velocity. We have to find the optimum blend of margin and velocity, the one that will generate the greatest amount of Throughput, i.e., the largest amount of Gross Income, given the reality of our playing field, given the parameters imposed by the market.

| Year | WIP | Closings | Cycle Time |
|------|-----|----------|-----------|
| 2005 | 100 | 225 | 160 |
| 2007 | 100 | 200 | 180 |
| 2008 | 100 | 240 | 150 |
|      | 80  | 240 | 120 |
|      | 100 | 300 | 120 |

"Back to the issue of variation and uncertainty", she said, gesturing towards the data table, and then looking toward the VP of Construction. "Earlier, you noted a widely-held belief, albeit intuitive and anecdotal, that RB Builders should be capable of building every house in less than 90 days.

"Not 180 days, not 150 days, not 120 days. In 90 days."

She wrote the following questions on the board:

*Where is variation being buffered?*
*What does variation cost?*

"The variation is the deviation from the standard. The standard is 120 days, and some of us think it should be 90 days. Considering the anecdotal evidence, your gut-instinct, and your interpretation of the data in the table, *how* and *where* do you think RB Builders' production system is buffering itself – protecting itself – from variation and uncertainty?"

"And – how much do you think all of this variation is costing RB Builders?", asked the intrepid, results-based consultant. "I will ask the question two different ways: How much is the lack of productivity costing you? How much is chronically-long cycle time costing you?

"Variation, productivity, cycle time; they are all connected."

"Our cycle time is 180 days, and we're doing it with 200 closings produced on 100 units of work-in-process", said the CFO. "Without getting into an explanation of statistics, we have about a 50% variation from the 120 day standard and a 100% variation from the 90 day duration we instinctively believe we should be achieving.

"Based on what you're saying, a system will protect itself from variation and uncertainty with some combination of longer cycle time, higher work-in-process, and excess/unused capacity. Our cycle time is long, which translates into a large buffer. Our work-in-process is where we planned it to be, and only 25% higher than the lowest amount it could possibly be, which is 80 units at a cycle time of 120 days. So, I do not think this is a case of RB Builders having a buffer of additional work-in-process.

"Here's the kicker:

"If we really believe we can produce 300 closings with our current production capacity, then we have a huge buffer of excess/unused capacity, which calculates into an overall utilization rate of 67%; in essence, we waste one-third of our capacity. As you have pointed out, that is hardly the picture of high productivity. The inability to more-fully utilize our production capacity translates into significantly fewer closings. We are paying for the capacity and the work-in-process to produce 300 closings, but we only closed 200 homes.

"'In terms of what all this variation is costing us, there is clearly a cost associated with excess work-in-process and unused production capacity", the CFO continued. "The additional, 'beyond-necessary' work-in-process certainly makes us a bigger company than we need to be, and the excess/unused production capacity alone costs us over $2,800,000 a year.

"But – I do not think it is about cost. I think it is about opportunity. Unless we opt for cost-cutting and reducing overhead – a 'same-for-less' proposition – then I would say that it is really costing us the opportunity of all the Gross Income on those 100 closings we missed in 2007. Our Gross Income Margin was 22%. We had $50 million in Revenue. If you divide that by the 200 closings that we achieved, the average sales price was $250,000. From there on out, the math is pretty simple."

The CFO walked up to the erasable board, and wrote:

$250,000 X 22% X 100 = $5,500,000

"We gave up $5,500,000 in Gross Income."

The conference room was quiet. Everyone was aware of the Gross Income Baseline, Target, and Reserve, and the impact an additional $5,500,000 would have on the payout of Gross Income Milestones under the new team-based performance compensation plan RB Builders had just enacted.

The silence was broken by the words of the CEO.

"No", he said. "That calculation doesn't even scratch the surface. What's the real cost?"

He walked to the front of the conference room.

"It's a whole lot worse", bristled the CEO. "If all that this excessive variation were to cost us was $5,500,000 in Gross Income, that would be bad enough. But − since we are profitable, despite the current market conditions and our own performance − it's also $5,500,000 that would have dropped straight to our bottom-line, in the form of additional Net Income.

"In terms of the cost of our production capacity, utilizing it would have cost us nothing − zip, nada, zero. We already paid for it.

"I'm not finished with this issue."

The CEO turned to the erasable board, picked up a marker, and added two more rows to the data table. Specifying a cycle time of 90 days, he purposely left the WIP and Closing data with a question mark.

Turning and gesturing toward the data table on the board, he said:

"Forget the 180 day cycle time we have now. What do you think RB Builders would look like at a cycle time of 90 days, instead of the 120 days specified under our job schedules?

"Don't bother. It's a rhetorical question."

The CEO made a quick calculation, and filled in the Closing and WIP data that he had initially excluded from the table.

"We could go with virtually any combination − and strategy or tactic − of higher Throughput and lower work-in-process to achieve a 90 day average cycle time, but let's say that we elect to go with a tactic of 'Max-T', a tactic of generating maximum Throughput with a planned, finite, and controlled level of work-in-process.

"Forget 200 closings on WIP of 100 units, which would be the current cycle time of 180 days. Forget 300 closings on WIP of 100 units, which would be the cycle time specified in our construction schedules. What about 90 days? Although it is a valid option, set aside the idea of making RB Builders a smaller company − 240 closings, but reducing WIP to only 60 units.

"Focus on Max-T. What about 400 closings on WIP of 100 units?

"What happens?"

| Year | WIP | Closings | Cycle Time |
|------|-----|----------|------------|
| 2005 | 100 | 225 | 160 |
| 2007 | 100 | 200 | 180 |
| 2008 | 100 | 240 | 150 |
|      | 80  | 240 | 120 |
|      | 100 | 300 | 120 |
|      | 60  | 240 | 90 |
|      | 100 | 400 | 90 |

The CFO handled the question. "Well, if our Gross Income Margin remained the same – which it likely couldn't and wouldn't, because of the price elasticity of supply and demand – we would generate an additional $11 million in Gross Income, every penny of which, as you point out, would drop straight to our bottom-line in the form of additional Net Income.

"Which means that our unwillingness – or, our inability – to do anything about the current level of variation that drives our long cycle times is costing RB Builders as much as $11 million in Net Income every year.

"That's a lot of money."

"Yes", said the CEO. "Even for a company that would then have $100 million in Revenue. I want everyone to be clear. Our 2008 baseline is $50 million in Revenue, from which we expect to produce $11 million in Gross income and $2,500,000 in Net Income. Our 2008 target is $60 million, producing $12.5 million in Gross Income and $3.4 million in Net Income. As all of you realize, all of our very-considerable bonuses are tied to the Gross Income Reserve that represents the difference between the baseline and the target.

"By appearance, this kind of performance would be over-the-top. However, if we pull it off, it means that, instead of progressively splitting a GI Reserve of $1,500,000 three-ways between our owners, Retained Earnings, and all of us, dropping our cycle time from 180 days to 90 days means we get to split a GI Reserve of $12.5 million.

"And, you know what? Despite the fact that its Revenue would have doubled, RB Builders would be the same size company. It would have the same amount of work-in-process; it would have the same amount of overhead, i.e., the same level of Operating Expense; it would have the same working capital requirement; it would have the same level of debt."

The VP of Sales weighed in on the discussion. "Rather than ask where your collective heads just disappeared to, let me just say that what you just described is not the reality with which we are dealing. This coming year – 2008 – is going to be a challenge. So, what happens when we are faced with the prospect of both fewer sales and lower margins? What happens when the market is not going to allow us to use – to economically leverage, as you like to put it – all of this new-found production capacity?

"I agree that productivity and throughput-killing variation is the problem when we are faced with an internal constraint, when we are faced with being our own worst enemy. Under those circumstances, 'Max-T' is the right approach.

"But, what happens when we are faced with an external constraint?"

"Give us some idea of what you are talking about", said the intrepid, results-based consultant. "Demand is elastic. How many sales, at what margin?"

"The numbers in the GI Baseline and GI Target are a couple of months old", the VP of Sales replied. "They already reflect the expectation of a deteriorating market. We still might be able to achieve those numbers, but, since then, the market has deteriorated even further. It is precipitous. We are looking at the very real possibility of both fewer homebuyers and lower margins."

"How few and how low?", asked one of the sales representatives.

"In 2007, we closed 200 homes, but we only sold 180 homes", said the VP of Sales. "Not exactly the protective backlog of sales that we want. So, we were already seeing the pressure building in the market. In 2007, our Revenue was $50 million, our average selling price was $250,000, and our Gross Income Margin was 22%.

"We kept those numbers in the baseline for 2008. However, the 2008 target is $60 million in Revenue, produced on 250 closings, with an average sales price of $240,000, and a Gross Income Margin of 21%. The higher productivity, higher utilization – whatever you want to call it – actually budgets more Gross Income in 2008 than we earned in 2007.

"Now, however, we are thinking maybe only 140 to 150 sales. We think that the average sales price will still be $240,000, but there will be more concessions. More concessions will result in lower margins, somewhere between 18% and 19%. If that scenario happens, we are looking at Revenue of $34.5 million, and Gross Income of $6.5 million.

"In case you missed it, our indirect, non-variable cost is budgeted at $8.5 million.

"That is an operating loss of $2 million, and, frankly, we don't know where the bottom of this recession is. It could get much worse."

"Nevertheless, we need to talk about what we know now. What would you do?", the intrepid, results-based consultant asked the VP of Sales. Glancing toward the CEO, she said, silently with her eyes, "Just let him answer."

"We can't afford to lose money. But – I can't look at the people in this room, and suggest that we fire people whom we have developed and whom we care about, either", said the VP of Sales. "Certainly not as our first resort.

"The market may not turn out to be this bad, but here's what I would do:

"The current job schedule says that we should be able to build our houses in an average of 120 days, and we have been given 100 units of work-in-process to produce as many closings as we can", he said. "That calculates to 300 annual closings.

"Despite the challenges of the market, we need to find a way to get those 300 closings.

"Let me qualify part of that statement.

"The closings are important, but we need to produce as much Gross Income as we can, because that's all we get to keep from whatever Revenue we generate from those 300 closings.

"From a production standpoint, someone else will need to figure out how to beat 60 days out of the current cycle time. From a sales and marketing standpoint, to sell 300 homes, I believe we will need to drop our prices to an average of $230,000. I know. It's a difficult decision. It's $10,000 below the target, and $20,000 below the baseline.

"Our margins would suffer, dropping to 15%, on average. We would need to become much more intuitive and instinctive in our adjustments, and learn to make decisions as fast and as frequently as necessary. We will have to fight for every sale. But – even with the lower margins – if we manage to sell, build, and close 300 homes, our Gross Income would be $10.4 million, produced on Revenue of $69 million.

"Our Net Income Margin would be less than 3%, but we would be profitable.

"Of course, like I said, it could get worse.

"Much worse.

"Could we find ways to extract more value, and therefore earn higher Gross Income Margins and generate additional Gross Income on every dollar of Revenue? I think so. Could we get our cycle times down to 90 days, and close 400 homes on the same amount of production capacity? Maybe. That would depend on us. I'm not sure that is what we would want to do right now, because the resulting higher production would have even further implications on prices and margins.

"Still, I'm starting to realize some things: First – higher velocity can overcome lower margins. Second – productivity gains are permanent. And, so are the benefits."

"In 2008, higher productivity is a case of business survival. And, it may not be sufficient, if things get a lot worse", said the intrepid, results-based consultant, writing as she spoke. "In the future, the ability to produce more – more closings, more Revenue, more Gross Income – with a finite and controlled amount of work-in-process and overhead will be one of the keys to RB Builders' sustainable competitive separation."

*Breakeven Analysis*

"This discussion raises a question", she said, pointing to the board. "From both a production standpoint and a financial standpoint, at what point does RB Builders breakeven?

"Knowing the answer to that question gives you more insight than you can imagine."

"One of the advantages and benefits of allocating costs on the basis of how they behave in relation to changes in Revenue is the ability to understand and use breakeven analysis", replied the CFO. "But, I have to admit, before you came, we could not have answered that question. The RB Builders Income Statement was prepared according to the NAHB Chart of Accounts, which is to say that it was comparative, compliant . . . and utterly useless. Now, we also produce a Contribution Income Statement.

"According to the 2008 baseline budget, our breakeven point is 155 closings, based on Revenue of about $39 million. However, because of the way the market has deteriorated, the 2008 target budget has a higher production breakeven point; it requires closer to 170 closings, albeit on only slightly higher Revenue of $40.5 million."

"How did you calculate that?", asked the VP of Construction.

"Let me show you", said the CFO, creating a new data table and adding the data to the first two columns.

|  | '08 Base | '08 Target | '08 Worse | '08 Full |
|---|---|---|---|---|
| Closings | 200 | 250 | 140-150 | 300 |
| Revenue | 50,000,000 | 60,000,000 | 34,500,000 | 69,000,000 |
| SP | 250,000 | 240,000 | 240,000 | 230,000 |
| GM | 22% | 21% | 18-19% | 15% |
| Breakeven | 155 | 170 |  |  |
| Breakeven | 39,000,000 | 40,500,000 |  |  |

"This is what we have said, so far. Some parts we don't have yet.

"Breakeven occurs at the point where overhead is completely absorbed", he said, continuing to write as he spoke. "Overhead is absorbed by generating Gross Income, which is the proceeds that we get to keep from each closing. When you're dealing with averages, one way to figure the breakeven point is to take the average sales price of a home, multiply it by the Gross Margin Ratio, and then divide the resulting Gross Income per home into your overhead."

Breakeven = OE ÷ (avg. SP x GM ratio)
Breakeven = $8,500,000 ÷ ($250,000 X 22%) = 155

"That gives you the unit breakeven point, in other words, the breakeven point in terms of closings. The unit breakeven point in the 2008 Baseline is 155 closings."

"Is there another way to look at breakeven?", asked the intrepid, results-based consultant.

"Sure", he said, writing on the board. "You can calculate the breakeven point in terms of Revenue. You calculate that by dividing overhead by the Gross Margin Ratio, which – as we all now understand – is basically the same measure as Contribution Margin.

"Take the 2008 Baseline and Target we were just discussing."

$$\text{Breakeven} = OE \div GM \text{ ratio}$$
$$\text{Baseline Breakeven} = \$8{,}500{,}000 \div .22 = \$38{,}636{,}000$$
$$\text{Target Breakeven} = \$8{,}500{,}000 \div .21 = \$40{,}476{,}000$$

"Like I said, about $39 million and $40.5 million, respectively", he said, pointing back to the data table.

"Our overhead under both the 2008 Baseline or the 2008 Target is $8,500,000, but the resulting Gross Margins are different, so the breakeven points are different. In this case, the difference in the unit breakeven point is more substantial than the difference in the Revenue breakeven point.

"Overhead is the same thing as Operating Expense, which is comprised of all our indirect, non-variable costs. Overhead – or Operating Expense – is the closest representation of the cost of our production capacity.

"That gives you the breakeven number of closings, and the breakeven Revenue. Equally-important – since we can't generate all of our closings and all of our revenue at once – is the breakeven rate.

"In terms of the scenarios we've been discussing, breakeven in our worst case scenario occurs at around 190 closings, which is at about $46 million in Revenue. That's because we're trying to absorb the same amount of overhead with smaller Gross Margins. It requires more closings, which can become a vicious cycle, with a lot of margin pressure.

"In the worst case scenario, the point at which we fully absorb our overhead occurs at a rate of about 16 closings per month. Of course, we have the production capacity to start and close 25 houses a month."

The CFO filled in the missing data for the last two columns.

| | '08 Base | '08 Target | '08 Worse | '08 Full |
|---|---|---|---|---|
| Closings | 200 | 250 | 140-150 | 300 |
| Revenue | 50,000,000 | 60,000,000 | 34,500,000 | 69,000,000 |
| SP | 250,000 | 240,000 | 240,000 | 230,000 |
| GM | 22% | 21% | 18-19% | 15% |
| Breakeven | 155 | 170 | 190 | 246 |
| Breakeven | 39,000,000 | 40,500,000 | 46,000,000 | 56,600,000 |

"Obviously, at only 140 to 150 closings – which amounts to only 12 closings per month – and $34.5 million in Revenue, we would be below the breakeven point in both closings and Revenue, which means we would be losing money.

"But, if we can somehow find a way to more fully-utilize our production capacity – which we will pay to have anyway, unless we cut our overhead – and somehow find a way to close more homes, albeit at considerably lower margins, we would breakeven at 246 closings and $56.6 million in Revenue. That occurs at 20-21 closings a month, a rate also well below the 25 closings-per-month we have the capacity to produce.

"At that level of productivity and capacity utilization, we have the ability to sustain even more diminished margins, if that became necessary. That's definitely not where we want to be on the margin situation, but it clearly demonstrates the competitive advantage higher productivity and utilization gives us."

"While we are talking about breakeven rates, I hope it reinforces the importance of even-flow production", said the intrepid, results-based consulting, looking at the VP of Construction and VP of Sales. "You cannot be all over the map each month with sales, starts, and closings, and with resulting WIP."

"Is there any kind of housing market in which demand does not respond to lower sales prices?", asked a sales representative. "What happens if we reduce the sales price – give more concessions, endure lower margins – and there are insufficient sales? Is there a limit to how far we can drop prices?"

"There's always that possibility", said the VP of Sales. "But, we still have to price to the market. That decision is not in our control. To the extent that we extract value, with better margins as the outcome, we do have some control. On the other hand, the gains from higher productivity should be permanent, and the speed/velocity that enables those gains is something we can control."

The intrepid, results-based consultant turned to the sales representative.

"To answer your question, if the situation gets bad enough, higher productivity becomes moot", she said. "RB Builders would have excess – and probably unusable – capacity.

"If we can't make some combination of higher margin and higher velocity work for us, we might have to take RB Builders out of gear, and glide to some sort of safe landing; a controlled crash would be a better description. This would be the equivalent of something like the situation NASA faced on the Apollo 13 mission; forget getting to the moon, just try to get the *Odyssey* and her crew home."

"Perhaps", said the CEO. "But, in that case, I would still want to come out of this mess with competitive separation. I'll be damned if I will accept that much pain without having something to show for it. Higher productivity might not seem as urgent, right now, but I would want to emerge a much faster and more agile homebuilding company.

"Forget 'industry best practices'; we need to do better than that, because we would be foolish to believe that our competition will forever come from who it comes from now. Someday, the homebuilding industry is going to change, and that change will as likely come from without, as from within."

*"Ex Disastrium, Scientia"*, said the intrepid, results-based consultant, smiling at her adaptation of NASA trivia. "Learn from adversity, learn from failure, learn from mistakes.

"Don't just learn from it; do something with it."

"There's one other issue we have to consider, but it's a separate one", said the CFO. "Under our new approach to team-based performance compensation, the cost of the bonus plan is intended to be completely self-funding, born from increases in Gross Income above the GI Baseline.

"Under the worst case scenario, there would be a small, unintended payout from the early milestones, but what that really points to is the importance of having baselines and targets that reflect current reality."

"I have one more thought", said the VP of Sales. "The last thing we want to do is to resign ourselves to treating our houses as just a commodity. We need to resist that temptation. If that's what this has become, then we simply aren't doing a good job marketing, on either the upstream side or the downstream side. We can remedy that, but it won't happen overnight. In the meantime, we may need to reduce our margins, in order to simply survive this down market."

"Okay, that was a lot to cover on variation and uncertainty", said the intrepid, results-based consultant. "There is much more to learn, and we will come back to variation and uncertainty in a bit. Before we move on and talk about scheduling, I want to make one more point.

"We never talk about variation and uncertainty outside of the context of a system. It's not just the presence of variation. It's the effect – the destabilization – that variation has on the dependent nature of the parts within a system. Variation anywhere in a system affects every part of a system. It's variation plus dependency that creates problems."

She moved back to the board, underlining the second discussion point.

"Let's shift the discussion to the second element and talk about how we schedule production processes. As we said earlier, these two elements play off one another, with scheduling being the key to managing variation."

1. Variation/Uncertainty
2. Scheduling

"For those of you who thought the last part sounded like physics, just wait. This part is really going to be physics – production physics. It is going to seem very theoretical, and I am not even giving all of it to you. You are going to have to trust me and accept it as the laws that apply to production. Not blindly. With an open mind. I need you to grasp the basic concepts, and just stick with me. I promise I will give you something concrete.

"But – you simply cannot understand what we are doing, unless you understand the concept. That's the way it is."

"Thanks for being so understanding", said the CEO.

"Don't worry, I'm not", replied the intrepid, results-based consultant, as she continued writing on the board.

## Lean Homebuilding

"The principles of Lean Production are to (1) precisely specify product or service value from the customer's perspective, (2) identify the value stream for each product or service, and eliminate as much of the non-value-adding steps/work as possible, (3) make the remaining value-adding work flow without interruption, (4) pull work into the system at the rate of customer demand instead of pushing it into the system, and (5) pursue perfection, which is really continuous improvement.

"From an enterprise standpoint, RB Builders is deciding whether it wants to embrace the tenets of Lean Homebuilding – and, if so, how it wants to embrace those tenets – when it comes to standardization of work, elimination of waste, visual management, kaizen, PDCA problem-solving, A-3 planning and policy deployment, and other areas.

"Culturally-speaking, RB Builders has to decide whether it wants to become a Lean Homebuilding enterprise, and thereby embody the most useful and transferable elements of Lean Thinking. In RB Builders' case, Lean Homebuilding's most beneficial contribution, to this point, is having awakened the need for embedding a process of continuous improvement.

"However – Lean Production, even crafted as Lean Homebuilding, is not the total answer. By itself, Lean Production – or Lean Homebuilding – cannot get RB Builders to where it needs to go.

"As useful, beneficial, and vital as Lean Homebuilding has been in the area of continuous improvement, it has not been as effective in the area of production management, in the areas of what is generally known as 'flow', at least, not straight out-of-the-box.

"Like every other worthwhile production method, Lean Production doesn't come from a homebuilding environment. That is a problem, because – as a homebuilding company – RB's production system has different parameters, faces different conditions, and imposes different requirements.

"We could discuss this for hours, but we don't have the time", she said. "Let me give you several quick examples to highlight the type of 'Lean issue' we see in various areas of production:

"First of all, Lean Production places a heavy emphasis on what is termed 'Just-in-Time' replenishment – JIT – which is the principle of producing only what is needed/ordered, leveling demand, leveling production, pull, continuous flow, etc.  But – how does that work in homebuilding?

"As the second example, consider production leveling, or heijunka.  A typical Lean manufacturer levels production based on forecast orders, not actual customer orders.  Some companies are better at making and adjusting forecasts than others, but, at best, it's a mix of 'change-to-order' and 'build-to-order'.  In a lot of cases, it's really 'build-to-forecast'.  To the extent there is variation in the forecasts, companies either have to carry a large inventory of finished goods, have to promise very long delivery dates, or have to live with a lot of excess/unused capacity.

"That would be the equivalent of RB Builders having to very accurately forecast the demand for every plan it offered in every community – or – live with some combination of an enormous inventory of completed homes (in addition to its required work-in-process), long delivery date promises, or a ton of unused production capacity.

"How in the world does something like that work in homebuilding?

"Finally", she said,  "consider the challenge of achieving continuous flow with a totally outsourced labor force, in a fragmented value stream, with as many manufacturing facilities or production plants as we have communities.

"How would we make that work?

"Think about it."

"So – are you saying that we should not consider a commitment to Lean Homebuilding?", asked the VP of Construction.

"No.  That's not what she's saying", answered the CEO.  "She's saying find a way to use the tools that work best for us, without regard to the religion or the denomination from which they came.  Understand the playing field.  Understand the parameters.  Understand the world we live in.

"Do what works.  Above all, get results."

He wrote it on the board.

Do what works
Get results

"I'll admit, at first, I didn't understand it", he said. "Or, at least, I didn't fully appreciate it. I had a hard time accepting the proposition that all the methods of Lean Production – in particular, the methods dealing with scheduling and flow – didn't automatically translate into something we could use. After all, if Toyota could do it, why couldn't we?

"What I have since come to understand is this: *Our production system is a blend of methodologies, but, it is also – by design and by necessity – a unique, proprietary expression of how RB Builders plans and manages production from the standpoint of what it is – a homebuilding enterprise.*

"Our production system is part of what differentiates us from other builders, and creates a sustainable competitive separation. That's what we want. We may not be there yet. But – that's where we are going.

"What we do – from the standpoint of production management in a homebuilding enterprise – combines elements of Lean, Six Sigma, Theory of Constraints, and other methods, but doesn't mirror any one of them", the CEO continued. "It takes different methodologies and makes sense of them from a homebuilding perspective."

"I'm sorry", sighed a superintendent, looking at the CEO. "Please explain to me why we need to know production physics. Why don't you just tell us what to do? You've obviously spent more time at this than us. We would be content to get our start packages, manage our jobsites on a daily basis, and leave the understanding of production physics to someone else."

"Fair enough", replied the CEO. "As an enterprise, we haven't always been as diligent or interested as we should have been in getting those of you who actually do the work – contract it, schedule it, inspect it, approve it – to also design the work and solve problems. There is no shortage of justification for empowering teammates to make decisions and get results. The emphasis on results-focused, team-based performance compensation should be evidence of that.

"But – restricting the answer to the topic of discussion – RB Builders' production system calls for more than managing jobsites on a daily basis, and assuring quality construction.

"It also calls for planning and managing production at the community level, through all three community life-stages, which includes paying attention to upstream and downstream marketing, pricing, flow, capacity utilization, maximizing throughput, controlling WIP, allocating resources – basically, everything required to plan and manage homebuilding production at the community level.

"Superintendents have to manage the schedule for each of their jobs on a daily basis, even though we also have to manage all of the schedules for all of the jobs as part of a portfolio, at both the community and company level. Recall what we said about systems. The parts have dependent relationships, and what affects one affects all the others.

"So, you have to do your part", said the CEO. "And – you can't do it, unless you understand it. This deep knowledge and understanding of production principles and disciplines has to become second-nature to you, a rapid, instinctive, and intuitive response to the conditions that you see, same as many other areas of RB Builders' production system.

"As leaders, we can't spend all of our time telling you how to do it. That's not continuous improvement. Principles and disciplines don't change, but our deep knowledge and understanding of them – and our ability to effectively apply them – improve continuously over time. This ongoing process of continuous improvement is as much a part of your job, as anything else."

Nodding in agreement to the CEO, the intrepid, results-based consultant continued, waiting until she had everyone's attention. Then, she added a couple lines to the CEO's reminders.

Do what works
Get results
Learn the concepts
Adapt/apply the concepts to homebuilding

"What I am about to say is very important", she said. "We are going to use a lot of Lean terminology, and some other terminology. Don't get distracted or hung-up on the terms that we use. From a production management standpoint – from the standpoint of how you manage a production system – all of these methods have terms that can be difficult to apply in a homebuilding environment.

"We have defined the terms that are important in RB Builders' production system. Learn the concepts. Adapt and apply that understanding in the context of our system."

The intrepid, results-based consultant continued, making another list on the board and underlining the first numbered issue.

Pace
1. Protection
2. System

"Every method of production planning and management involves the scheduling of a sequence of tasks performed by resources", she said. "Central to that understanding are three issues, framed by these questions: What sets the pace of production? How do we protect the process from variation and uncertainty? How do we manage a process as a system, and not just a collection of independent activities?

"Let's consider the first process scheduling issue. What should we be using to set the pace of a production process?"

"Doesn't the job schedule set the production pace?", asked a superintendent.

"Not really", argued a different superintendent. "The job schedule sets the sequence and expected durations of activities for each house  There's sequence and duration, but no rhythm. That's what pace is – the rhythm of the production system. Pace applies to the entire system, not each individual job. If we don't see it as a production system, we just have a collection of separate, disconnected job schedules."

"I would think pace has to do with a resource."

"Okay", said the intrepid results-based consultant. "Presume the pace of production should be set by some type resource. What kind of resource? What kind of attributes would you look for in a resource tasked with setting the production pace of the entire system?"

The ideas came in rapid succession from different members of the team.

"The most important resource?"
"The most expensive?"
"The most reliable?"
"The busiest?"

"Lean Production tries to schedule a single point – the pacemaker – to set the pace for the entire system, process, or value stream", she said. "The pacemaker sets that production pace to match the rate of customer demand, or what Lean calls 'takt'. The purpose of 'takt time' is to precisely match production with customer demand. It is the heartbeat of a Lean Production system.[6]

"As you just noted, the pace of production is what gives the production system its rhythm", the intrepid, results-based consultant continued. "At least, pace should give production its rhythm; we've seen the consequences of RB Builders' failure to achieve even-flow production. What did you call it, a tsunami?

"Speaking of which, a term related to 'pace' is what we call 'flow'."

On the board, she wrote:

## Continuous Flow and Build-to-Order Processes

"In the Lean World, there are different types of flow. The production mantra in Lean says, 'flow where you can, pull where you must'. Lean is very much a continuous, or single-piece, flow proposition, with a recognition that continuous flow is not always possible. As such, Lean rejects old-fashioned, batch-and-queue, mass production; it encourages continuous, single-piece flow as its picture of perfection, and has tended, in the past, to look at production management in a factory or manufacturing environment.

"Homebuilding is essentially a build-to-order proposition", she said. "From a production process standpoint, ordering and flow are separate considerations: How do you sequence orders? How do you make production flow?. The nature of build-to-order manufacturing creates challenges to achieving the continuous, single-piece flow that Lean Production would

---

[6] 'Value Stream Mapping', Lean Enterprise Institute, 2000.

prefer. But, since Lean will accommodate build-to-order processes, it should be able to work in a homebuilding environment.

"For a build-to-order process, Lean recommends maintaining a FIFO, or First-In, First-Out, sequence, regulating the amount of inventory or work-in-process, and making the bottleneck resource function as the pacemaker.[7] Since continuous flow and build-to-order processes are both pull-type production systems, the distinction lies in where you put the pacemaker in the process.

"In continuous, single-piece flow, Lean puts the pacemaker as close to the end of the process, as close to the customer, as it possibly can", she said. "Everything upstream from the pacemaker can be pulled – replenished – from small amounts of inventory, at the demand of each downstream activity, and then everything downstream from the pacemaker is continuous flow.

"In build-to-order processes, Lean puts the pacemaker earlier – upstream – in the process, at the point from which FIFO sequencing begins", she continued. "That applies directly to homebuilding, because – in addition to being a build-to-order process – homebuilding should use FIFO sequencing. Lean also recommends making the most constrained resource in a build-to-order process the pacemaker, all of which makes it possible to consider pacemaker placement based on parameters and requirements other than customer demand and continuous flow.

"Which brings the discussion to something called the Theory of Constraints", said the intrepid, results-based consultant, adding to a previous list.

*Lean Homebuilding*
*Theory of Constraints (TOC)*

"Despite their differences – which I don't want to diminish – all improvement approaches and methodologies do overlap each other, from the standpoint of the elements they share, the elements they have in common", she said. "And – educated bunch that you have become – you say, well, duh, the physics that govern production are immutable, and the various improvement approaches and methods should have more in common than they have in conflict.

"Reducing an improvement approach or methodology to its salient features is an over-simplification. Nevertheless, essence can be useful. Every approach or methodology has its salient feature, which – conveniently enough – tends to identify it.

"So – we say that TQM is about quality; we say that reengineering is about radical redesign; we say that Six Sigma is focused on reducing variation; we say that Lean is about reducing waste.

"The salient feature of the Theory of Constraints – what we call TOC – is identifying, exploiting, and elevating the system's constraint to higher throughput. We have already discussed constraints. Recall the earlier discussion of chains; the weakest link in the chain

---

[7] 'Creating Continuous Flow', Lean Enterprise institute, 2001

invariably limits the strength of a chain – limits its ability to accomplish its purpose; that's the constraint.

"No matter what else you improve about the chain – no matter which other link you strengthen – you will not improve the overall strength of the chain, unless you strengthen its weakest link. It is the constraint that determines the throughput of the entire system.

"Think about the implication of that last statement", said the intrepid, results-based consultant.

"What the Theory of Constraints gives you is focus. Not a focus on something you don't like and believe is inherently wrong – or inherently evil, as former GE Chairman and CEO Jack Welch would say about variation – and about the methods for how to reduce or eliminate it wherever it exists.

"No.

"TOC gives you a focus on what you have to improve in order to make more money.

"In terms of production scheduling, Lean and TOC actually have a lot in common, and they both have a lot to offer. Most of their purported conflicts are overstated, particularly within the narrow confines of homebuilding. The areas of actual conflict between Lean and TOC are few. In terms of scheduling, the most significant difference between TOC and Lean is that TOC purposely unbalances the capacity of the resources of its process, while Lean purposely levels the capacity of the resources of its process.

"Let me show you what I mean", she said, writing on the board.

*Continuous Flow and Build-to-Order Processes*
*- v. Synchronous Flow*

"TOC builds production scheduling around the availability of the system's most constrained resource, because it is this constraint that determines the throughput/output of the entire production system", she explained. "TOC is all about finite capacity, and synchronizing the entire production system around whatever makes the capacity finite, what it calls the constraint. Lean likes continuous flow; TOC likes what it calls synchronous flow.

"Lean acknowledges the presence of constraints, what it terms bottlenecks, but doesn't require its pacemaker to be the constrained resource – unless, it's a build-to-order process. When it comes to how they manage their build-to-order processes, the Theory of Constraints and Lean Production are in essential agreement.

"And – homebuilding production is what type of process?"

"It's a build-to-order process", said the CEO.

"Very good", said the intrepid, results-based consultant. "That's the refrain in this little hymn. Or, maybe it's a responsive reading. I'm not sure. Anyway, whenever you hear the question, that's the answer.

"Let's give it a try", she said, as she wrote on the board.

*Drum-Buffer-Rope*

"Theory of Constraints uses what it calls Drum-Buffer-Rope (or DBR) to simply convey the concept of scheduling production according to the capacity of a specific resource", she said. "Whatever TOC calls its scheduling method is not important. Drum-Buffer-Rope is simply the terminology that TOC uses to describe a way of scheduling a production process. Depending on the type of flow that it wants to create, Lean Production uses a combination of a pacemaker, takt time, pull, kanbans, FIFO, and inventory to achieve essentially the same idea.

"In either case, that same idea involves a process that (1) schedules its production according to a selected pace, (2) agrees on the resource the system will use to maintain that pace, (3) ties the work of the remainder of the resources to the pace-setting resource, and then (4) protects the entire system from variation and uncertainty.

"Lean schedules its pacemaker to meet customer demand or forecast, while TOC schedules more to maximize Throughput, by fully-utilizing the capacity on the most constrained resource – on the bottleneck or the constraint. Other than that, TOC and Lean essentially agree on this point.

"There is one significant difference", she said. "TOC purposely unbalances the capacity of the system, in favor of managing a planned or imposed constraint, and subordinates all of the other non-constraint activities/resources to it. Lean Production purposely levels, or balances, the capacity of the system by means of heijunka (production leveling), demand leveling, and continuous flow – unless, of course, it's a build-to-order process, in which case, TOC and Lean at least partially agree, because Lean recommends selecting the bottleneck (or the constraint) as the pacemaker.

"And – homebuilding is what type of process?"

"It's a build-to-order process."

"Under DBR production scheduling, the 'drum' is the resource that sets the pace of production, and every other resource is subordinated to the requirements and needs of the drum resource. Under TOC, the drum resource is always the constraint, which means that it is the resource that determines the throughput of the entire system. Previously, we talked about a system being like a chain; the constraint resource is the chain's – the system's – weakest link. It is the activity/resource that either has the least capacity, has the greatest number of dependencies, or has the most work.

"The pace-setting resource in Lean Production is called the pacemaker", she explained. "For the most part, Lean doesn't much like constraints, and views them as something to be prevented, fixed, or avoided, not something to be managed. Lean prefers to level production and match it to takt-time.

"So, in Lean, the pacemaker is usually *not* the constraint – unless, of course, it's a build-to-order process. In which case, TOC and Lean essentially agree, because Lean recommends selecting the bottleneck resource – its constraint – as the pacemaker.
"And – homebuilding is what type of process?"

"It's a build-to-order process."

"That's right", said the intrepid, results-based consultant. "Under either TOC or Lean, the entire production schedule is tied to the pace set by this single activity/resource – to the pace set by this pacemaker/drum – whether that resource is the constraint or not.

"Under Theory of Constraints DBR, the mechanism that ties the schedule to the drum is called the 'rope'. The rope is the synchronizing mechanism, and the outcome is synchronized flow. In Lean, the schedule depends on the type of flow – on whether we have continuous flow, pull or FIFO. If we have continuous flow, release is not an issue and we don't need a release mechanism; if we have pull, the mechanism is something that acts like a kanban.

"As you recall, Lean would prefer continuous flow – 'flow where you can, pull where you must'. If Lean cannot create flow, it will resort to pull; it just has to have something to pull with."

"Like a rope?", quipped a superintendent.

"Lean and TOC are both known as pull systems, in part, because of the actions of the rope and kanbans", she continued. "For a number of reasons, RB Builders very much intends its production system to be a pull system.

"We will talk more about pull systems later in this session.

"TOC and Lean both use buffers to protect their processes from variation", she explained. "As we learned from the Law of Variability Buffering, unless we do something to reduce the variation, every – and, I mean every – production process will find some way to buffer or protect itself from the effects of fluctuation, using some combination of reserve/excess capacity, time, or inventory.

"The difference is in how the buffers in DBR and Lean work. In TOC, the various types of buffers protect the capacity of the constraint. In Lean, the buffer is dispersed more widely, because the capacity has been leveled."

The CFO raised his hand. "Go ahead", she said.

"Something you just said triggered a thought", he said. "When we were talking awhile back about how we are currently buffering for variation and uncertainty – and how much that variation and uncertainty is costing us – I made the observation that we seemed to have large time and capacity buffers, but not a particularly large inventory buffer."

"That's right. RB Builders doesn't have a large inventory buffer", said the intrepid, results-based consultant. "Why? Why do you think that is?"

"I have no idea", he said.

"In part, it is because high levels of work-in-process won't protect a build-to-order process from variation", explained the intrepid, results-based consultant. "As long as starts are pulled into the system at the rate of closings, the FIFO sequencing generally holds down excess inventory, because each resource can only work in a prescribed, first-in-first-out sequence.

"If homebuilding production was a continuous or single-piece flow process, it could more easily protect itself with any combination of the three buffers, but it's not continuous or single-piece flow. As we know, homebuilding is a build-to-order process, and a build-to-order process generally buffers itself with excess capacity and extra time, not with excess inventory."

"I agree. We don't currently have an inventory buffer", replied the CFO. "But, at times, we have, in fact, had high levels of work-in-process, and . . . "

" . . . you want to know why that happens", she said, smiling. "Simple. A build-to-order process can certainly contain excessive levels of inventory, but that inventory is not functioning as a buffer that protects the system from variation, because it's not a case of the system responding to variation. Excess inventory either happens – or doesn't happen – as a result of the internal policies that allow it or prevent it. It's detrimental, in the same way all excessive inventory is detrimental. In the case of a build-to-order process, higher inventory doesn't protect the system from variation. To the contrary, excessive inventory exacerbates the problem, and makes managing production and utilizing production capacity even more difficult.

"It's muda."

The chant began in the back of the conference room, low at first, but growing in volume and intensity, until it engulfed the entire room.

"Mu-da! Mu-da! Mu-da!"

"All of which is a nice little segue into what we have to talk about next", said the intrepid, results-based consultant, pointing to the list. "Process Scheduling Issue No. 2 is about how we protect a production process from variation and uncertainty."

1. Pace
2. Protection
3. System

"Didn't we just cover variation and uncertainty?", asked a sales representative.

"Yes – we did", she said. "However, we discussed it from the standpoint of the pacesetting feature of scheduling a production process – from the standpoint of the drum-constraint resource of TOC and the pacemaker resource of Lean. We said there were three issues. The first issue was pace. The second issue is protection. We know what the process tries to protect, and we know what the process tries to protect itself from."

"Now, we just need to close the loop.

"In any event, the presence of variation and uncertainty in every discussion we have had should tell you just how connected all of these principles and disciplines are, and how important it is to understand the effect of variation and uncertainty on the performance of a process and how it is managed."

She searched the board and flipchart, found the particular list she was after, and added another term to the top.

Variation
Buffers:
1. Higher WIP
2. Excess/Unused Capacity
3. Longer Duration

"When we talk about the Law of Variability Buffering, it is not just about buffering and protection", she continued. "It is about the parameters – the measures – that frame the performance of the system. It is about the relationship between four measures: variation, capacity, time, and inventory. It is about the real world choices we have to make about how much variation we want, how much variation we will tolerate, and – then – how we will deal with it.

"These measures cannot be considered in isolation. We cannot decide to reduce cycle time and not consider how that decision will affect capacity and inventory, particularly absent any commitment to reduce variation. Buffers have a cost. A system doesn't protect itself for free. As I have previously explained, if we don't reduce process variation, we will bear its cost in the form of some combination of longer cycle time, unused capacity, or higher levels of inventory. The problem is not the buffer; the problem is the variation that causes it.

"First of all, there are two categories of buffers", she said. "That is, categories of buffers, not types of buffers; it's an important distinction, category versus type of buffer. The first category is known as a 'strategic buffer'. It is purposely left in the system for product mix, choice, fluctuation in customer demand, whatever. The second category is known as a 'surplus buffer'. Its very name should tell you that it isn't good. It's muda.

"Let's look at a scenario. What do you think would happen, if RB Builders, one day, decided to mandate a reduction in the level of work-in-process, from 100 homes, down to 80 homes, without reducing the amount of variation in the system?", she asked.

"Those kinds of things – imperial mandates – never happen, right?"

"Oh, of course not, not at RB Builders", said the CFO. "But – if it happened to – the system would respond by demanding more capacity and longer durations. In other words, it would increase the capacity and time buffers."

"Okay", she said. "But, what would happen if RB Builders also mandated that its capacity can't change, that the company was prevented from incurring any increases in operating expenses or resources? Same amount of variation, less inventory, same capacity.

"What would happen then?"

"Well, in that case, we would have smaller inventory and capacity buffers, so the system's only protection from the same amount of variation would be the time buffer, and – as a result – the cycle time on every house would go through the roof", said a sales representative.

"Based on the equations you gave us for calculating Little's Law and the Law of Variability Buffering, I would say our average cycle time would go from around 180 days to about 225 days."

The intrepid, results-based consultant rummaged through a previous flipchart and found the formulas she wanted.

$$CT = 120 \text{ days} \quad WIP = 80 \text{ homes} \quad T = 240 \text{ homes}$$

$$WIP = (CT \times T) \div Days$$
$$(120 \times 240) \div 360 = 80 \text{ homes under construction}$$

$$CT = (WIP \div T) \times Days$$
$$(80 \div 240) \times 360 = 120 \text{ days}$$

$$T = (WIP \div CT) \times Days$$
$$(80 \div 120) \times 360 = 240 \text{ closings}$$

"Understand the connection", she said. "If RB Builders wants to be fast, but it doesn't tackle variation, then it needs to resign itself to having a lot of excess/unused capacity to buffer it. On the other hand, if RB Builders wants low levels of work-in-process, but doesn't address variation, then it better be prepared to endure long durations or reduced output."

"I have a couple of questions", said a superintendent. "The concept of variation buffering makes sense, but how does a production system sort all this out? You can say that you won't allow any buffering. Is that even possible?"

"No, it's not", replied the intrepid, results-based consultant, answering the second question first. "It doesn't matter if you say you won't allow buffers. A system will buffer variation. The Law of Variability Buffering is no different that the Law of Gravity. The question is only how it will buffer the variation.

"It is easier to take away the inventory and capacity buffer", she continued. "You can cap inventory or freeze hiring, and probably make it stick. But, time is really a self-determining buffer. Duration is the buffer overflow valve. Absent some progress on reducing variation and some combination of higher inventory or excess capacity, you will simply wind up mandating cycle times that cannot possibly be achieved.

"There are no free lunches.

"Production performance hinges on these four interdependent measures: variation, time, capacity, and inventory."

"Let's talk about the third process scheduling issue", said the intrepid, results-based consultant, pointing again to the list. "Let's talk about how the process is managed as a system."

1. Pace
2. Protection
3. System

"Earlier, we were talking about the concept of Drum-Buffer-Rope as a 'picture' of how a process is scheduled", she said. "DBR is a TOC term, but the concept of DBR differs very little from the concept of scheduling contained in Lean Production, in the sense that (1) both methods have a pace-setting resource, (2) both methods tie the schedule to that pace-setting resource, and (3) both methods protect the process from variation and uncertainty.

"Both approaches involve a process that schedules its production according to a selected pace, agrees on the resource it will use to maintain that pace, ties the work of the remainder of the resources to the pace-setting resource, and uses buffers to protect the entire system from variation and uncertainty.

"It should be very clear that none of this works, unless we manage it as a system", she said. "Our view of how our homebuilding production system works is not the sum of the efficiency or effectiveness of a collection of related-but-unconnected parts. It is the effectiveness of all the parts working interdependently toward a common goal.

"So – how does all of this play out in RB Builders' production system? Given the same parameters – the same playing field – that apply to homebuilding, Lean Production, the Theory of Constraints, and every other reputable approach to managing production in accordance with 'factory physics' – the physics that apply to production management – would agree on basic points.

"These points all come together in RB Builders' production system", said the intrepid, results-based consultant.

"Production is planned and managed at the community level, throughout the three life-stages of every community. We have the ability to shift our capacity to meet demand and maximize Throughput, but we have a planned, finite, and controlled amount of work-in-process and a planned, controlled, and finite amount of production capacity with which to do it.

"The ideal is even-flow production – an even and sufficient rate of closings, that pulls an even and sufficient rate of starts into the system, buffered by a protective number of sales", she said. "However – as we know – even-flow is an outcome, not a mechanism. RB Builders cannot mandate even-flow, it has to produce even-flow.

"We produce even-flow by managing the system as a system", she said. "We set the proper production pace, we synchronize everything to that pace, and then we protect the system from the effects of 'Murphy'.

"We beat variation and waste out of the system, but we still protect the output of the system from the inevitable delays that occur whenever something goes wrong, from anytime it is impacted by the variation and uncertainty that remains, from the errors, delays, and waste that remain."

The intrepid, results-based consultant moved back to erasable board and added two sub-points to the third scheduling issue.

1. Pace
2. Protection
3. System
    a. Pull v. Push
    b. Balanced Capacity v. Unbalanced Capacity

"RB Builders' production system is a pipeline, with two control mechanisms. It is controlled by both the rate of closings and the production capacity and rate of the resource selected as the 'drum' – the constraint resource or pacemaker. The two control mechanisms are synchronized, so that RB Builders produces the 'close one, start one' pull that is our mental model of even-flow.

"The drum is 'roped' to a start matrix, which acts as the kanban/gate, releasing available starts into the system", she explained. "The function of the start matrix is simply to set the priority of starts, not the rate of starts. The release of every start is triggered by the actions of the control mechanisms, first by a closing, and, then, by the available capacity on the constraint resource and the amount of existing work-in-process in front of it.

"RB Builders has selectively placed internal 'buffers' that protect the output of the process from variation and uncertainty, with a combination of reserve-protective capacity on non-constraint resources, and, perhaps, with a small amount of work-in-process in front of the drum (constraint/pacemaker) resource. The entire building process is protected by a buffer of available starts, which buffers the start matrix from fluctuation in sales with an inventory of starts."

"I have a question", said a superintendent. "You've been harping on productivity, but now you're adding capacity? Why?"

"Harping. It's the wings, right? They're a dead giveaway", answered the intrepid, results-based consultant.

"We are not adding capacity, we are permitting it to exist. By definition, every resource other than the constraint or the pacemaker has excess capacity. It goes back to our understanding of systems.

"In the Cost World, this excess capacity would be considered waste, and it would be eliminated. But, in the Throughput World, excess capacity on non-constraint, non-pacesetter resources is considered reserve capacity that protects the throughput of the system from variation and uncertainty, by making certain the pacesetter works at capacity. When it is the constraint, the pacesetter determines the throughput of the entire system.

"Again – which resource does Lean Production recommend as the pacesetter-pacemaker in a build-to-order process?"

"The constraint", answered the superintendent.

"And – what type of process is homebuilding?", she asked.

"Ooooh, it's a build-to-order process", sang the two sales representatives. "Build-to-order, build-to-order. First-In-First-Out, FIFO, don't you know. Ooooh – oooh!"

"Nice", said the intrepid, results-based consultant, smiling.

"From a process 'flow' standpoint, when RB Builders' production system is in an even-flow state, it can maintain the FIFO sequencing that is required for a build-to-order process. It only expedites when individual jobs are in danger of finishing late and missing their completion date."

The intrepid, results-based consultant moved to the flipchart and flipped back to a previous page.

*Variation/Uncertainty*
*Pull Systems*
*Unbalance Capacity to Balance Flow*
*Scheduling*

"Okay", she said. "Remember the original list we created. First, we covered variation. Then, we covered scheduling. The other two subjects – pull systems and balanced capacity – we said fit under scheduling, but they need to be discussed.

"First – let's talk about pull systems. A 'pull' production system produces only in response to downstream 'customer demand'. While the adjacent downstream activity is an internal customer – and important in a pure manufacturing environment – the customer demand that

RB Builders is concerned with is the demand from its ultimate customer – which are its homebuyers, in the form of presales.

"On the other hand, a 'push' system produces as much as it can, without regard to downstream internal customer demand, or even external market demand", she continued. "In a manufacturing environment, 'push' – and batch processing or 'batching' – often happens in order to satisfy local efficiency and utilization measures.

"However, in a homebuilding company, push occurs more out of a false notion of production disciplines, or from trying to meet quarterly or annual Revenue targets."

The CEO stood and spoke.

"Our production system is a pull system, but that wasn't always the case", he said. "The current system – which is new – replaced a push system. Like the current system, the old system had a start matrix, too, but it controlled both the rate and order of starts in a community. The starts slotted into the start matrix were regarded as untouchable – once they were scheduled, starts were not supposed to be missed under any circumstance.

"It was noted that the entire building process is protected by a buffer of available starts, which protects the start matrix from fluctuation in sales with an 'inventory' of starts. That's what we call the start buffer, and it actually sits in front of the start-to-completion process.

"Every other homebuilder proudly calls this their 'sales backlog', as if having a six-month backlog of contracts that they can't start is something they actually want. Under the old system, we looked at the backlog that way, too – the longer the backlog, the better. We would sell a house in April, with no intention of starting it until October. Now, we take a different approach.

"If our controls will allow the start, we certainly don't want to miss it, but – over time – we concluded that a start buffer with a 60-day backlog provides sufficient protection for the start matrix, and corresponds to the current length of the contract-to-start process. Any lengthier backlog – a start buffer that's any larger – is muda; too much to manage, too much that can change. Our homebuyers hate the wait, and it really is counter-productive.

"Under our old system, the matrix did a good job of producing starts at an even rate, and gave needed order to what was previously a totally chaotic process, but pushing starts into the system without regard to the throughput – the rate of closings – resulted in higher levels of work-in-process. Absent any increase in either productivity or production capacity, RB Builders' cycle times would lengthen, to the point of sometimes reducing the rate of closings."

The intrepid, results-based consultant picked up where the CEO left off.

"That's a good explanation", she said. "For most of us, once we've had a chance to think about it, 'pull' makes immediate sense, because it is linked to the demand that justifies the production. But, the issue of capacity – whether it should be balanced or unbalanced – is less intuitive.

"It is a paradox.  It is a paradox that a production system with balanced capacity cannot achieve an even rate of production.  The term 'balanced capacity' means a production system that has its capacity distributed evenly throughout the system", she explained.  "It means a production system designed with the same capacity at every resource.

"A production system with the same production capacity at every resource – in other words, a system that purposely levels its capacity across all of its resources – cannot have even-flow production.  Intuitively, we believe the opposite, that a 'balanced' system – one in which resources have the same capacity – produces balanced results.  However, in a production system with balanced capacity, variation and uncertainty *anywhere* in the system will affect production *everywhere* in the system, making it impossible to control or predict.

"Moreover, production systems with balanced capacity tend to be very rigid and difficult to manage.  They are not the adaptable, agile, easily-managed system we want RB Builders' production system to be.

"Even-flow production – which we want – is an outcome, not a mechanism.  In order to have an even rate of production, we have to purposely unbalance the system that produces it, and create production 'pull' instead of production 'push'.  We have to live with some amount of excess or reserve capacity on the non-constraint, on the non-pacemaker resources."

"We touched on it earlier in this session, but unbalancing the system flies in the face of Lean and TPS", said the VP of Construction.  "If we want to embrace Lean Homebuilding, don't we have to do heijunka?  Don't we have to do production leveling?"

"In a controlled manufacturing environment with continuous flow, production leveling works", said the intrepid, results-based consultant.  "The plant can move equipment, cross-train workers to do other workers' jobs, shift production cells, change product mixes, adjust production runs, etc.  Problems can be solved much more rapidly in that environment.

"However, in an environment like homebuilding, heijunka is exponentially more difficult to achieve.  Homebuilding is not a controlled manufacturing environment.  It is the equivalent of building cars in people's driveway.  Instead of teammates, we have trade partners that are independent sub-contractors and suppliers, who also work for our competition.  We deliver materials to hundreds of jobsites.  The more inherent variation and uncertainty is to a process, the more difficult it becomes to level production.

"Moreover, homebuilding is not the continuous, single-piece flow process that Lean prefers.  It is a build-to-order process", she said, motioning for the SR Chorus to keep their seats.  "Lean gives you its first clue about the feasibility of production leveling in a build-to-order process, by recommending that the most constrained resource – the bottleneck – become the pacemaker.  If the capacity and production rate of every resource is leveled to the capacity and production rate of the constraint, then, theoretically, there is no constraint.  And – theoretically – you would have unlimited capacity.

"Yet, we know systems do not have unlimited capacity.  As we have noted, there is always a constraint.  There is always a weakest link in the chain.  We are far better served purposefully placing the constraint and subordinating everything else to it, than we are fighting for a system with balanced capacity.

"This has been an incredibly long and difficult session", said the intrepid, results-based consultant. "Unquestionably, a vitally important session, as well. There is more that came out of this session – more that you learned from this session – than you can possibly imagine. Having an understanding of how production processes are managed is critically important.

"You did a great job."

# CHAPTER VI:  MANAGING PROJECTS

"Hard to believe, but we are getting close to the end of our series of workshops on the production principles and disciplines that define homebuilding", said the intrepid, results-based consultant.  "We have two more sessions.

"Thus far, we have covered the visual image of production – that conveyed by a pipeline.  We have talked about how operating performance is linked to business outcomes, through the decisions we make every day as a team that affect Throughput, Inventory, and Operating Expense.  We have learned about systems – how they behave as sets of interdependent parts that must work together to achieve a common goal, and the way of reasoning, thinking, and solving problems based on that behavior.  We have talked about how homebuilding production is managed from a process perspective.

"Now – it is time to talk about project management, specifically as it pertains to homebuilding."

She sorted through the dog-eared flipchart that was always used to chronicle the topics of discussion before those discussion points disappeared from the erasable board.  She found the sheet she wanted.

Production System = repository of processes and projects
Processes = end-to-end steps, input into output, internal perspective
Procedures  = short, simple, one person, no handoffs
Processes = longer, more complex, cross-functional
Value Streams = higher-level, external perspective, info, data, kaizen
Project Portfolio = Managing projects in a multi-project environment

"So – what is a project?", she asked, underlining as she spoke.

"I was doing a bit of research on this topic", said the CEO.  "The International Standards Organization (ISO) defines a project as 'a set of coordinated activities, with a specific start and finish, pursuing a specific goal, with constraints on time, cost, and resources'.  I think that's a pretty good definition.

"The true nature, the essence, of homebuilding – the actual process of building homes – is project portfolio management, not process management or value stream management", he said.  "Homebuilding is project management, performed in a multi-project environment.

"I will say this:  I think we have to understand project management in the context of process management;  we have to understand process management in the context of project management;  and we have to understand production in the context of both.

81

"As an enterprise, RB Builders is a group of people doing things.  The *things* we do can be classified as either process or project.  A production system has both processes and projects.  In a sense, projects can contain processes and processes can be comprised of projects, but they behave differently and have different attributes.

"And – they are managed differently", he said.

"The nature of *processes* is that they are continuous, end-to-end sets of activities that transform inputs into outputs", said the intrepid, results-based consultant, picking up where the CEO had left off.  "Processes can be simple or complex, they can be functional or cross-functional, they can be transformational or transactional, but they tend to be done the same way each time[8].  There is a defined, standardized set of business processes that defines RB Builders' workflow, like its contract-to-start process.

"The nature of *projects*, on the other hand, is that they are longer and more complex.  Projects have defined beginning and ending dates, and defined interim milestones.  Projects tend to require more outside resources, and there tends to be more resource contention; project workflow is less continuous, and projects are more schedule-intense", she said.

"The more a set of activities has to be modified, the more new – or different – the ground that it covers, the more likely it is to be classified as a project.  The more continuous and repeatable a group of activities is, the more likely it is to be classified as a process.

"The scheduling requirements associated with projects – including how we resolve resource contention – is the biggest difference between project management and process management.

"In process management, the entire process is scheduled, not the individual job or specific input being transformed.  In project portfolio management, each individual job is scheduled as part of the project portfolio;  the release of new projects into the project portfolio is known as 'pipelining'.

"Of all the sets of end-to-end activities that RB Builders could construe as a process, the start-to-completion process behaves most like managing a project portfolio, like managing multiple projects."

She pulled a previous sheet from the flipchart.

Critical Path
Logistics
Projects, Processes, Value Streams:  What do we schedule?

---

[8] Process experts can – and do – stake out different positions on the spectrum between rigidity and flexibility.  It is a matter of finding the best way to deliver value;  it is a matter that must weigh benefit and cost;  it must determine what is necessary.  Homebuilding does not have the same flexibility requirements as other industries might;  homebuilding is clearly not a case that necessitates ad hoc workflow.

"Remember", she said. "Processes don't have a critical path of activities like projects have. The logistics of how houses are built, the environment in which houses are built, and who does the actual work – that is about project management, not process management.

"Resolving resource contention – the same resource being required on different jobs at the same time – is much more a part of the project portfolio environment than it is the process environment.

"We schedule jobs, not the process. Jobs are projects. Our work-in-process is a network of projects.

"So – I have a question", she said. "Why do projects have such long durations – and why do they finish late?" She looked at the Vice President of Construction. "You explained the symptoms of the problem in the previous session. If you wouldn't mind, give it to us again.

The VP of Construction nodded his willingness, but warned, "Everyone needs to understand that these are simply my observations. I think these observations relate to the problem, but I don't have the solution.

"The problem starts with variation and uncertainty", he said. "Variation is related to what we call Murphy's Law, which simply says that if something can go wrong, it will go wrong, probably at the worst possible time. There are different kinds of variation. Some variation is common-cause, some is special-cause. Some variation is attributable to a specific cause, some is not. Some variation we treat with kaizen and PDCA, some we use risk management.

"The point is this: Variation exists, and it is a problem", he continued. "Variation and uncertainty lies at the heart of every other production principle, because it affects everything. What we do every day is largely in response to our need to control variation and deal with the risk of uncertainty: How we protect the system – buffer the system – from it."

"Variation – what we call common cause variation – is a matter of being either able or unable to achieve repeatable results", said the intrepid, results-based consultant. "Uncertainty – what we term special cause variation – is more a matter of not knowing what those results will be."

"That's right", said the VP of Construction. "That gets into the behavioral aspect of human nature, how we respond to variation. I agree with your earlier point, that natural human behavior is learned or acquired, but the response is so predictable, it might as well be natural."

He walked to the front of the conference room, pulled out the flipchart he wanted.

*Padded Durations*
*Student Syndrome*
*Multi-tasking*
*Parkinson's Law*
*Murphy*

"So – what is this natural human behavior we are talking about?

"First of all, people tend to increase the estimated duration of tasks, in order to provide more certainty and higher reliability. They tend to pad their estimates of task duration. It's understandable. They want to get their work completed on schedule. They don't want to be late. Take our framing contractor as an example. It should take him 10 days, but he tells us 20 days.

"It's not meant to be deceptive, but it's still hidden.

"Then, the other tendencies of human behavior conspire to consume all of that hidden safety. People wait until the last minute to start, until they have used up the safety. In the meantime, they look at the schedule, see the extra days, and they work on other things. The first behavior is known as Student Syndrome; the second is known as multi-tasking.

"In addition to Student Syndrome and multitasking, there is a general tendency to allow every task to expand to whatever time that it is given, something known as Parkinson's Law. So – even if we specify an early start, the duration of the task tends to expand to whatever completion date it's given.

"As a result of these behaviors – Student Syndrome, multi-tasking, Parkinson's Law – delays accumulate, but early completions are lost. Then Murphy shows up, and the job finishes late. Or, we finish that particular job on time, but the effect of expediting it ripples through all the other jobs.

"Here's the deal: If you add up the hidden safety built into every task, across an entire job schedule, we are looking at an enormous amount of padding – extra time – in our schedules."

"How much padding?", asked the CEO.

"Well – our typical schedule is 120 days, but we all know we should be able to build on a 90 day schedule", replied the VP of Construction. "So, if that's any indication, I would say the safety is 30 days, which is 25% of the schedule. But, apparently, it's not doing its job.

"The padded durations intended to let us complete homes more consistently – albeit with what is clearly a longer-than-necessary schedule – appears to be wasted, because the jobs wind up taking upwards of 180 days."

"I understand", the intrepid, results-based consultant said. "Let's take a moment to see where we are. So far, we have talked about what projects are, where they fit in the big scheme of a production system, and the behaviors that cause them to finish late, or, at best, take longer to complete than they should.

"As the projects comprising a project portfolio, this thinking would clearly apply to jobs in a homebuilding production system. In fact, you should just go ahead and substitute the term 'job' for 'project'. For our purposes, the two terms are completely interchangeable.

"The production component of homebuilding is clearly about project portfolio management, about managing projects in a multi-project environment", she said, adding, "I think you will find that the discussion of projects from here-on-out is going to be much more intuitive and instinctive for you.

"It's all connected, and we will keep coming back to all of these points.  I want to finish the discussion on duration, and then talk about how projects are structured, how they are scheduled, and how they are managed."  The intrepid, results-based consultant made a list on the erasable board.

*Project Duration*
*Project Structure*
*Project Scheduling*
*Project Management*

"We have heard that jobs take too long and usually finish late.  Rest assured, you are not alone.  Every company in every industry vertical struggles with this same issue.  However, duration and missed deadlines are just the symptom of the problem.

"We also heard about some of the human behavior that is driven by our lack of understanding of variation."  As she spoke, she dug through the previous sessions' flipcharts until she found what she wanted.

*What is the core problem?*
*What is the solution?*
*How do we implement it?*

"What we have to figure out is how to fix it, how to solve the problem.  We have to answer questions:  What is the core problem?  What is the simple, elegant solution?  How do we implement the change?

"What to change, what to change to, and how to make the change.

"Any ideas?"

One of the superintendents spoke.  "Well, why not get our sub-contractors to accept shorter task durations?  Just tell them that's the way it has to be.  Based on what we're saying, they don't need all of the time they are given in the schedule."

"They don't use the time they are given productively, either", added another superintendent.

"No, the work doesn't take as much time as they are given, and they don't utilize the time they are given as effectively as they should", said the intrepid, results-based consultant.  "Some of that ineffectiveness is clearly their fault.  But, some of it is our fault, and the rest is human nature.  And, if we arbitrarily mandate shorter task duration, we are simply going to accept the illusion of shorter task duration, and substitute that illusion for the reality of higher project completion risk.

"Duration is the symptom of the problem. What is at the root of longer-than-necessary task duration?"

"Longer duration is one of the buffers that protects a system from variation and uncertainty, correct?", asked the CFO. "Then, the next question becomes, what are we trying to protect? We are protecting the completion date of the task, right? And, the presumption is that protecting the completion date of each task protects the overall completion date of the job.

"Something tells me that padded duration spread between all of the tasks doesn't protect the overall completion date of the job, but, nevertheless, the safety – the padded duration – on each task is there because of both."

"Because of both what?", asked a sales representative.

"Because of both the presence of variation and uncertainty, and the need for dependable completion dates", said the CFO.

"Go on", said the intrepid, results-based consultant.

"It occurs to me that part of the problem is in the confidence level of the durations we request, and that our sub-contractors offer", said the CFO. "I think we both want the safety, which is why we can't capriciously set shorter durations. I think we sub-consciously drive expectations for highly-confident schedules, and I think that our sub-contractors respond to those expectations with task durations they are 90% to 95% confident in meeting.

"Highly-confident task durations supposedly make achieving job schedules certain, but it also makes them long, much longer than the actual time required to complete the work. This buffering of variation and uncertainty is part of the problem, and it is compounded by the fact that it is hidden. It's not meant to be deceptive, it's just unspoken.

"The other part of the problem has to do with statistics and the laws of probability."

"That's why you like it when IBS is held in Las Vegas", said the VP of Sales. "It gives you an opportunity to test your theories of probability."

"I don't know about Las Vegas, but he's right about statistics and probability", the intrepid, results-based consultant said. "Take the standard deviation curve – a bell curve – and apply it to something like task duration, and you will find that the 'long tail' – extending the probability from 50% to 95% – increases the duration of a task by a factor of 1.64.

"The reciprocal of 1.64 is .6098, so you can figure it from either end.

"Take the earlier example of the framing sub-contractor. He may be thinking 10 days and telling us 20 days, but if he is 'very confident', 95% confident, that he can complete the work in 20 days, then, statistically, he has a 50/50 shot at completing the work in around 12 days.

"So, at a duration in which there is 50% confidence, there is a 50/50 chance that the framing will be completed in 12 days or less, and an equal chance that the framing will take 12 days or more", she said.

"Since 20 days is the 95% confidence duration, in theory, the framing should almost never take longer than 20 days.

"We know that probability going in, and we can plan for it.

"But, the way the framer, like everyone else, manages his work – waiting until the last moment to get serious about the work, using up all of the allotted time, working on multiple jobs – any of the time that could be saved framing the house tends to be lost, and that lost time accumulates on the schedule.

"All of the safety is wasted. It almost never takes less than 20 days to frame a house, and usually it takes more than 20 days."

The intrepid, results-based consultant quickly made a bulleted list:

Cascade effect:
- Task duration
- Job duration
- Community level
- Company level

"What we have been talking about are the problems that occur within tasks", she continued. "Extend the intersection of the tendencies of human behavior and statistics from the task level to the job level. The effect at the job level is much more than the sum of the effects at the task level.

"On the rare occurrence when a task finishes early, the resource that performs the next task is usually not ready, so the job waits. What happens if a task finishes late? The resource on the next task has to wait, so it multi-tasks other jobs, and they all finish late.

"Synchronization is lost, the dependency ripples throughout the job, and the delays 'cascade' – they accumulate – throughout the job.

"A few minutes ago, we heard that RB Builders' typical schedule is 120 days, the current cycle time for a job using that typical schedule is 180 days, and – regardless of what the schedule says – we should be able to routinely build to that typical schedule in 90 days.

"Presuming the 120 day job schedule reflects the sum of highly-confident task durations, about half the time, those jobs should be completed in 73 days; forget 90 days. And, the 180 day actual cycle time we experience is a picture of all the delays, on all the tasks, cascading through the entire job schedule.

"180 day cycle time is reality.

"These are the problems within each job", she said. "Extend the intersection of the tendencies human behavior and statistics from the job level to the portfolio level – from the job level to the community level, and then on to the company or enterprise level. The more 'things' that are in-process – the more tasks, the more jobs, the more communities the system has to contend with – the more complex and difficult the management dilemma becomes.

"The dependency ripples throughout the system, just like a pebble thrown into a lake. The delays accumulate throughout the system, and they cascade throughout the system's portfolio of projects."

"I want you to confirm something", said the VP of Sales.

"What's that?", she asked, flipping the capped erasable marker over her right shoulder and catching it behind her back, without looking at it, or ever taking her attention off the VP of Sales.

"All our lives and throughout our careers, we have been told that multi-tasking – the ability to work on more than one thing at a time – was to be considered a prized personal attribute. How many times have we heard it and seen it? You know – put it on your resume' – 'has the ability to multi-task'", he said. "But, you are saying it is not an attribute, that multi-tasking is to be avoided."

"To the extent that multi-tasking was ever construed to mean the act of juggling a set of equally-important tasks – the act of keeping a lot of things in the air, like a set of plates – then, yes, you were definitely misinformed", the intrepid, results-based consultant replied, continuing to flip the erasable marker over her shoulder and catch it behind her back, without ever dropping it, without ever looking at it, or diverting her attention from the VP of Sales.

"You know – juggling.

"We've said it before. Managing a production system requires concentration; it requires focus. Focus requires prioritization of tasks.

"If a task is the most important, critical step in a process – i.e., the step that should be the pacemaker, the constraint, the bottleneck, the task that determines the throughput of the entire system – then you cannot divert attention to a task that is not as critical", she said. "And – even if it is not the constraint, multi-tasking without focus and priorities is counter-productive."

She drew two simple charts.

"Look at the first chart", she said. "You have four projects in-process, which you choose to work on at the same time. You do them concurrently. You multi-task them. The tasks on each project requires one week to complete. When will the first project be completed?"

| | Week 1 | Week 2 | Week 3 | Week 4 |
|---|---|---|---|---|
| Project 1 | --- | --- | --- | --- |
| Project 2 | --- | --- | --- | --- |
| Project 3 | --- | --- | --- | --- |
| Project 4 | --- | --- | --- | --- |

"It will be completed in Week 4", said the VP of Sales. "All of the projects will finish in Week 4."

"So – does multi-tasking meet the requirements of homebuilding? Does this look like the way we want to schedule jobs? Is this even-flow?", she asked. "Let's look at the alternative, which is to complete a project – in its entirety, all of its tasks – before moving on to the next project."

| | Week 1 | Week 2 | Week 3 | Week 4 |
|---|---|---|---|---|
| Project 1 | ------------- | | | |
| Project 2 | | ------------- | | |
| Project 3 | | | ------------- | |
| Project 4 | | | | ------------- |

"Look at the second chart. Same deal, but you do the projects in sequence. You complete them one-at-a-time. You don't multi-task. In this case, when will the first project be completed?"

"It will be completed at the end of Week 1", replied the VP of Sales.

"In the first table, when does the first throughput occur?", she asked.

"Early in Week 4", said a superintendent.

"In the second table, when does the first throughput occur?", she continued.

"At the end of Week 1", he said.

"All of this work is being performed by Resource 1", she said, "Now, look at when the next resource, Resource 2, gets to work on a project. In the first table, Resource 2 gets all of the projects completed by Resource 1 essentially at the same time, at different points during Week 4. In the second table, Resource 2 gets Project 1 at the end of the first week.

"In both cases, it's the same amount of work. In either case, Resource 1 should finish all four projects by the end of Week 4, so we get the same amount of throughput from Resource 1 during the entire four-week period.

"But – we get the throughput much faster when Resource 1 does not multi-task its projects.

"If these tables represent systems, which system requires the most capacity? Which system carries the most work-in-process? Which one has the most complexity?"

"The one that kills more of my brain cells", answered a superintendent.

"Let's try this again", said the intrepid, results-based consultant. "Which scenario has more things that can go wrong? Case in point, in the first table, what happens to Projects 2, 3, and 4, when you have a delay on Project 1?

"Get my drift?"

"It's an illustration of the finite capacity and variability buffering we have been talking about", she said. "If you fix the level of capacity and increase the amount of inventory, and do nothing about variation and uncertainty, the system has no choice but to lengthen cycle time. That's the undesirable effect of multi-tasking, similar to making things in batches, which is very un-Lean. Multi-tasking and batching are not the same thing, but they have similar consequences.

"And, remember, multi-tasking is not the only behavioral tendency that causes the long durations we are concerned with", she continued. "We pad durations to create safety, but then we waste all that safety, by waiting until the last moment, by letting the work expand to the allowed time, and – yes – by working on multiple jobs at the same time."

"That might be the case in 'real' project management, but not here, not on my jobs, not on any of our jobs; I would argue not with any good homebuilding company", argued a superintendent. "My framers have full crews on each job. Same with the electricians, plumbers, masons, and everyone else."

"Bullshit", said the VP of Construction.

"I can't tell you how many times your framer leaves three guys on your job and sends the rest of the crew to get another job started", he said. "It happens all the time. It happened to you last week. The truss package showed up a week late on Lot 40. The same day, the trusses showed up on Lot 47 three days early.

"Two jobs with trusses on the ground, and one framing crew to fly them. Where's the crane? On Lot 47. Which job is behind schedule? Lot 40. So, why don't we send the crane to Lot 40? Because we can't. Lot 40 has a purchase order issued to another crane operator.

"A lot of times, it's not even their fault. We tell them to do something, because we can't seem to get our own act together on resolving the resource conflicts between open jobs.

"But – it's not one behavior. It's all of it – the multi-tasking, the waiting until the last minute, the pacing of work to consume the allotted time. There is variation in all its forms. Different outputs from repeated applications. There is risk and uncertainty. Things that are indefinite, indeterminate, and unknown. Weather. Delays. Mistakes. Failed inspections.

"Things that go wrong. M-M-M-Murphy.

"And – we step right into it.

"If you can't see this, then you will have to come up with a different explanation – a different excuse – for why your jobs average 180 days", he said, "when the schedule itself specifies 120 days, and everyone acknowledges that it should only take 90 days."

The intrepid, results-based consultant let the VP of Construction's words sink in before she continued. "Final point", she said. "What is the relationship between throughput, and profitability and economic return? How do you tie operating results to business outcomes?"

The CEO stood.

"Throughput with a capital T", he said. "Shorter task durations lead to shorter cycle times. Short cycle times lead to higher productivity. With higher productivity, we produce more, and we consume less doing it. When we produce more-with-less, we generate Gross Income at a faster rate. That's Throughput – with a capital T. When we do that, we generate cash faster, we produce higher levels of Net Income, we are more profitable, and we generate a much higher level of economic return, a much higher ROA.

"We make more money. Fiscally, we are more conservative. Operationally, we carve out a degree of competitive separation.

"We protect our livelihoods."


"Excuse me. May I say something?"

The question came from RB Builders' most experienced, capable, and respected superintendent, someone not known for his commentary. The conference room grew silent-but-fidgety.

"Sure", replied the CEO. "Tell us what's on your mind."

"Nice speech."

"Glad you enjoyed it", said the CEO.

The superintendent continued, thinking carefully about the words he would use.

"You know me", he said. "When I first heard about this arrangement, it made me want to throw up."

He glanced at his notes. "Let me see. Ah, yes, here it is: 'Partnering and learning, participating in the financial outcome with a consulting firm on a series of projects with short timeframes and targeted, focused results, in pursuit of an overall company goal'.

"Really?

"I have to admit, this stuff is always kind of intriguing, even the idea of having team-based performance compensation. However – that's as far as I would ever let it go. In terms of expectations regarding the outcome – I have always been skeptical, bordering on cynical.

"I still am.

"This sounds like so many other programs the company has embarked on. None of them have worked. Not the way they were supposed to. Over-promised? Badly-executed? Focused on the wrong thing? I can't say, but I doubt that this program will turn out any different. It will just take up a lot of our time, sitting in meetings like this, that ultimately lead nowhere.

"What was it that was said? 'Think globally, act locally'? I like the second half of that statement.

"My response to this stuff has been to put my head down and do my job. I have become good at doing that. As far as I have been concerned, this company consists of my job, nothing more. I have relationships, but that's not about my work. That's about friends and associates.

"I have argued in favor of individual performance compensation, not because I disliked team-based performance compensation, rather, because I knew I could achieve the performance that triggered the individual bonus, without having to depend on anyone or anything else.

"Give me a cycle time to achieve, and I will achieve it. Give me a quality standard to meet, and I will meet it. It doesn't matter what others do, it doesn't matter what RB Builders does. I still get paid.

"During the course of the past month, as we have been told that we have to change the way we understand and manage production, that we have to accept team-based performance compensation, I have been ready to quit my job, if these changes were enacted. I am not willing to sacrifice what I know I can do personally, for what I do not believe we can do as a team.

"As for my job, I know what it takes to do it. You can have the rest of it. I don't need it. I could care less about production systems.

"I am the production system.

"It isn't that I don't think this new stuff is the right approach. To the contrary, it makes compelling sense to me. It's the way we should be doing it. However, it was bound to fight a losing battle against what I consider to be a pervasive attitude of complacency, entitlement, laziness, and resignation. All of you so-called leaders – those of you taking down the six-figure salaries and failing to produce results? Well – it leaves me a little cold.

"From my perspective, RB Builders is a good company, and actually one of the better places to work. But – it lacks the willingness, capacity, and capability to change. All we ever wind up doing is tying ourselves in knots, all the while declaring the latest thing we need to change.

"In the end, my skepticism wins out. I am simply not going to be stupid enough to risk my own performance compensation on something in which I have no confidence.'

The superintendent looked directly at the CEO. "In the past, it didn't matter as much. But – like you – I don't think average performance defined by 'best practices' is going to cut it any longer. Whether we choose to call it an analogy, a concept, a metaphor, or a simile, I do think our production system is a pipeline. I do think we need to understand the connection between the decisions we make every day and the business outcomes that protect our livelihood. I do think we should focus on optimizing the performance of the overall system, instead of the performance of its individual parts. I agree that there is a distinction between processes and projects, between process management and project management, and that we need to understand both.

"More than anything else, I want to be part of something that is bigger and more important than me. We have talked about the need for a savvy, accountable, and motivated homebuilding team, comprised of savvy, accountable, and motivated teammates. I want to base my work on the 'want-to' attitude that it takes to own and run a homebuilding business, not just the 'how-to' mechanics of selling, starting, and building houses. I want to share accountability and responsibility for the decisions and the results. And – yes – I want to have a serious financial stake in the outcome we create together.

"In that regard, most homebuilding companies seem to be pretty much the same, which is to say, stuck in the past. I doubt I could improve my personal situation much by leaving, particularly in the current market.

"So, maybe it doesn't matter.

"But, that doesn't erase the past. It does not dissuade my concern with this company's willingness and capability to enact these changes.

"I'm the one who's saying it, but plenty of us are thinking it. This may be a good approach, but it doesn't mean it will work in this company, not the way it has operated in the past. Tell me how this is going to be different. Tell us how this situation is somehow different, and will work in this company."

"Your concern – your issue – doesn't have anything to do with this approach. Or any other approach, for that matter", said the intrepid, results-based consultant, directing her words to the superintendent. "Your issue doesn't have anything to do with production principles, or with production systems."

"Oh, I see. It has nothing to do with production principles or production systems", said the CEO, directing his words to the intrepid, results-based consultant. "That's really profound. Of course, it is much more foundational than that. This is about giving teammates a reason to care, and about inspiring the confidence to follow where we are leading this company.

"I will address each of your points", he said, turning his gaze towards the superintendent.

"The steps we have recently taken to create business savvy-ness, to impart decision-making accountability and responsibility, and to provide a real financial stake in the business outcome – to every single member of the team – are radical by industry standards. Collectively, they give all of us – and I mean all of us – reason to care deeply about improving the operating performance that drives that business outcome.

"Some of us are charged with an official responsibility of leadership, by virtue of our management positions in the company. Fair enough. I think that the leaders of an enterprise are there to serve, not be served. But, leadership is not fundamentally about the position that you hold, or do not hold. The way I choose to describe leadership, it is a matter of personal character and courage. It is about the credibility that flows from demonstrating integrity.

"It comes from trusting one another. It means sticking to agreements and keeping commitments. It means going beyond the expectations and responsibilities of your individual job It also means speaking and hearing the truth. Expressing candor, like this", he said, nodding to the senior superintendent, "comes with the territory.

"It means pursuing goals that are worth pursuing, even if those goals are difficult to achieve. No. I'll re-phrase that. It means pursuing goals that are worth pursuing, precisely because they are difficult to achieve.

"It comes from inspiring the type of optimism that sees every situation as we choose to make it, not as we are told it must be. Or, how some long-standing, so-called industry experts tell us it should be, if we're willing to settle for so-called 'industry best practices'. Somehow, I just do not see industry best practice guidelines as the path to competitive separation. The assessment of current reality, that underlies everything? That is where we are.

"The production principles we have instituted? Vetted, to get us to where we want to go. Team-based performance compensation? The focus on Gross Income? Same thing.

"This is the direction in which RB Builders is headed.

"For whatever reason in the past, such an understanding of leadership has been absent. At best, it has been an obligation and a title, for some, and an excuse for everyone else. It is a situation that predates me, but it is a situation that stops with me. Repairing the damage will not happen overnight. That's the thing with integrity and credibility. It's difficult to build and it's really easy to destroy. But – the situation is going to change, that I promise you.

"Based on past performance, I won't argue with your conclusion", said the CEO. "Nor will I blame you for being – as you put it – skeptical-bordering-on-cynical, about the prospect for change, the approach to production management we are committed to, or the requisite leadership I have just explained.

"The right amount of skepticism can be healthy; the possibility for pessimism is always there, but it has to be overcome, if anything worthwhile is to ever happen. But – I am going to make this clear, to all of you, to all of us – cynicism is pointless. Get over it.

"This is the direction in which RB Builders is headed, and your choice to continue your cynicism has a life expectancy of about 30 more seconds."

The intrepid, results-based consultant looked over at the senior superintendent, shrugged, and said, "Now, that – that was a speech.

"I do agree with your point", she said, turning back to the CEO. "But, you know, it is possible to fail, without being the least bit cynical about its prospects." Turning back to everyone in the room, she continued. "I would say this: Whether you are cynical or not, change is up to you. Whether any of this works or not, is up to each of you individually, all of you collectively.

"If I didn't believe RB Builders would make it work, I wouldn't be here, my firm wouldn't accept you as a client. We would not waste our time with you. After all – by virtue of how we are being compensated – my firm and I have a bigger dog in this fight than anyone else.

"When we started out on this little adventure, I told you that my consulting firm would be compensated on the same performance basis as everyone else", she reminded them. "I told you that there was no limit to the time and effort that my firm – and I personally – would expend to achieve the outcomes we targeted together. I told you that I would work hand-in-hand with you, and do whatever it takes to achieve those goals.

"I assured you that I would do whatever it took to foster the willingness and the capacity for change, create a sustainable capability for implementing the things that would continuously improve operating performance and business outcomes, increase innovation and learning, and make you less dependent on all of your consultants. I told you, from the standpoint of how credit was attributed, that I was content to remain in the background.

"Those were the assurances I gave you", she said. "In return, I sought and received assurances from you.

"You agreed that this was a true client-consultant partnership, and that – because my firm's compensation was completely results-based, of finite duration, and self-funding – my firm was assuming the higher level of risk. You agreed that this new, results-focused consulting arrangement we were undertaking provided ample incentive to everyone for taking action, making changes, and improving operating performance and business outcomes.

"I told you that I was as serious as a heart attack about getting results. I made it clear that I had no intention of wasting my firm's time and effort. I told you that you did not have to do everything I told you, but that you did have to come to terms with me, take action, make needed changes, and do whatever it took to achieve the targeted results. Although I have grown rather fond of you – most of you – I made it clear that, if there was no action, no change, no results, then – out of principle alone – heads needed to roll.

"I want to tell you a couple of stories.

"My younger sister played NCAA Division II soccer in college. As a junior and senior, she was her team's captain, so, her junior year, before the season began, a sports reporter asked for her prediction about the season, specifically, what kind of won-loss record she thought her team would have, as a measure of the team's success.

"My sister replied: 'I don't plan to lose any games'."

"When my dad was younger, he sailed a lot of offshore races. According to him, it could get dangerous, sailing offshore at night, in rough seas. He says that they all used to remind each other, only half-jokingly, I think, that – where they were – if their boat went down, no one was going to blame them if they drowned.

"But – they knew they would still be dead, if they just gave up.

"Trust me, no one outside of RB Builders cares whether you succeed or not. No one else cares whether you separate yourselves from your competition. No one else cares whether you keep your jobs. Nobody else cares about your livelihood or your future. Nothing new in that revelation. Back in the Age of Homebuilder Entitlement, nobody cared, either. It was just never an issue, because being good enough was good enough. Success is no longer such a foregone conclusion. No one cares, and no one is going to blame you.

"But – that doesn't change the outcome.

"I can just hear it now", said the intrepid, results-based consultant. "Poor things. What a great company RB Builders could have been. It was just too much for them to handle, housing's version of The Apocalypse. It's not their fault.

"Nope. Nobody's going to blame you, if you go out of business. But, that is just what you will be – out of business.

"Failure is not an option. Not for me. Not for any of you. We are not giving ourselves that choice."

"We need a break", said the CFO.


As everyone slowly reassembled from the break, the intrepid, results-based consultant proceeded. "Let's move on to the matter of how projects are structured", she said. "This part will be short, because the points on either side – project duration and project scheduling – are more important than project structure.

"The importance of project management extends beyond RB Builders' production system, beyond how the portfolio of jobs is managed. For most enterprises, project management is a fact of life. The start-up of a new community is a process, but it is structured, scheduled, and managed as a project. The conversion to a new operating system is a project.

"Project structure is important, because scheduling and managing projects in a multi-project environment – which is exactly what homebuilding involves – is different than it is for a single project."

*Project Duration*
*Project Structure*
*Project Scheduling*
*Project Management*

"When we talk about project structure, we are really talking about how projects are planned and put together.  There are project management terms that need to be related to the terms we commonly use in homebuilding.  In the case of a homebuilding company, the 'project plan' for an individual job/project is reflected in its overall schedule.   Like most builders, the scheduling module of RB's integrated operating system structures individual jobs on the basis of pre-designed templates.

"That schedule is comprised of tasks and groups of tasks, in what is known as the 'Work Breakdown Structure', or the 'WBS'.  Looking at it from either a community or a building company level, homebuilding production is essentially the management of multiple projects, known as a 'project portfolio'.

"PERT charts and Gantt charts are graphic tools that provide different views of how a job/project is structured and scheduled.  They are common to most scheduling apps.

"PERT stands for Project Evaluation Review Technique;  PERT charts show the structure of a job/project as a task network diagram with multi-point estimates of task effort, to allow for uncertainty in task durations.   Gantt charts provide a timeline view of a job/project with dependencies, but does not provide for uncertainty, unless it uses 'float'.

"In most scheduling software applications, Gantt charts are more common than PERT charts.

"For right now, that's all we need to know about project structure.  I want to move on to project scheduling."

*Project Duration*
*Project Structure*
*Project Scheduling*
*Project Management*

"We have a problem", said the VP of Construction.   "Throughout this production workshop, references have been made to protective buffers, to buffering the system from variation and uncertainty.

"The scheduling templates in our scheduling application do not provide for buffers.   Our scheduling application allows you to insert 'float' as a separate task, but these float-type tasks aren't written into any of the job templates.  Even if they were in our templates, 'float' sounds like longer, added duration, not a buffer mechanism.  Perhaps you are planning to talk about this under project scheduling, but I don't think what we have now will work."

"Yes – we are going to address the 'float/buffer' issue under project scheduling", the intrepid, results-based consultant answered. "No – you cannot substitute float for buffering, because float adds duration without increasing protection.

"Therefore – yes – you do have a problem.

"As I said earlier in this session, the scheduling requirements associated with project management – including the way we deal with resource contention – is the biggest difference between projects and processes.

"As you recall, in *process management*, the entire process is scheduled, not the specific objects flowing through it; in this case, those objects would be the individual jobs. In *project portfolio management*, each individual job *is* scheduled as part of the project portfolio; the release of new projects into the project portfolio is called 'pipelining'.

"The differences between scheduling processes and scheduling projects can also be seen in the concepts used by the different management approaches, in order to explain them.

"For example, one of the two approaches we discussed in an earlier session on process management – Theory of Constraints – uses the concept of Drum-Buffer-Rope (DBR) to convey the idea of scheduling a process according to a specified pace. Under DBR, you select the resource – the *drum* – to set and maintain that pace, you create a mechanism – the *rope* – to tie the work of the remainder of the resources to the pace-setting resource, and you insert a device – a *buffer* – to protect the entire system – the process – from variation and uncertainty.

"DBR is a TOC concept, but Lean uses a concept with similar elements.

"Lean does not speak to project management to the degree it speaks to process management. Theory of Constraints speaks to both. There is no fundamental difference between a Lean approach to project management and a TOC approach to project management.

"However, there are significant differences between the Lean/TOC approach to scheduling and traditional project management", she said, walking to the flipchart and making another list. "There are also important differences between how you schedule a single project, and how you schedule a portfolio of projects. And – there is some other stuff you need to know.

"Here is what we need to understand about scheduling projects."

*Traditional (Critical Path) PM v. Lean/TOC (Critical Chain) PM*
*Single project v. multi-project*
*Task Duration (reducing) and Hidden Safety (removing)*
*Buffers*

"Traditional project management is a well-defined discipline", she continued. "It has its own institute – the Project Management Institute, or PMI – and a published viewpoint, the Project

Management Book of Knowledge, PMBOK for short. In regard to project scheduling, there were no meaningful changes/advances in that body of knowledge, nor in project management thinking, for more than 50 years – until the emergence of Critical Chain Project Management in 1997, which is the method upon which Lean/TOC project management is based.

"Both traditional project management and Lean/TOC determine the longest set of dependent tasks through a task network. One calls it a 'path' and the other calls it a 'chain'. The aggregate durations for this path/chain represents the fastest overall time in which a project can theoretically be completed.

"In traditional PM – that is, in traditional project management – this set of tasks is known as the 'critical path'; it uses the term Critical Path Method – or CPM – and Critical Path Project Management – or CPPM – to convey that idea. Lean/TOC goes a step beyond CPM, by taking the critical path and accounting – in other words, adjusting the schedule – for resource contention; resource contention occurs whenever the same resource is demanded on *different tasks in the same project* at the same time, or demanded on the *same task in different projects* at the same time.

"The first situation – *different tasks in the same project at the same time* – is about scheduling a single project", she explained. "The second situation – *same task in different projects at the same time* – is about scheduling a multi-project portfolio. Homebuilding obviously involves both situations.

"Someone give me an example of what I am talking about."

"When the framing subcontractor has four jobs at the framing stage, but only three crews", replied a superintendent.

"Yes, that's a good example", said the intrepid, results-based consultant. "We will get more into how it accomplishes that feat later, but Lean/TOC resolves any resource scheduling conflicts. The critical path with resource contention considered/resolved is what Lean/TOC calls the 'critical chain', using the term Critical Chain Project Management, or CCPM.

"CCPM usually lengthens the critical path, which is only a bad thing, if you are prone to suspending reality.

"The first difference you must understand between critical path and critical chain is that the critical path method only specifies the relationship and dependency between tasks, whereas the critical chain method specifies both task dependency and resource dependency.

"The difference between CPM-CPPM and CCPM, quite literally – is the difference between the characteristics of a path and the characteristics for a chain. They are, therefore, good descriptive terms."

"Well, if that's the case, then I think we have yet another problem", said the VP of Construction. "Our scheduling software uses the Critical Path Method (CPM), not critical chain."

"That doesn't surprise me in the least", she said. "In fact, I would expect it. The algorithms CPM scheduling programs use are based on thinking that is more than 50 years old. The scheduling software based on these algorithms is like a rotary dial telephone – except that it's still around.

'It's a coelacanth.

"CCPM is somewhat new, and the PM fraternity is characteristically slow to adapt changes and improvements. There are, however, ways to allow or adapt critical path-based programs to critical chain scheduling. There are a few standalone CCPM applications, but, to my knowledge, the scheduling modules of existing integrated homebuilding software packages are exclusively critical path method. In order for any builder – not just RB Builders, but any builder – to schedule using critical chain, it will require either an add-in application or a rewrite of the scheduling module in the existing system."

The intrepid, results-based consultant set the erasable marker in the tray.

"I see the look on your faces", she said. "No one is saying this is easy. The easy way is to adopt 'industry best practices', to settle. Our first instinct is to play it safe, to stay away from the edge, to take a middle-of-the-road approach.

"My clients in Texas have a saying: 'The only things in the middle of the road are yellow lines and dead armadillos'. My instinct reflects that observation. It is to be distinctive on the product side, and to be faster and more agile on the production side. There is no sanctuary in being average.

"I can just hear it now: 'Wow! RB Builders' production practices are on par with the best homebuilders in the industry!' As if the homebuilding industry had anything to write about, when it comes to adopting cutting-edge production practices.

"Why stop there?", she asked. "Everything we have been dealing with in these sessions – from the mental model of 'production is a pipeline', to making the connection between operating performance and business outcomes, to understanding how costs behave, to having a perspective and a discipline regarding the systems and processes that create the value we are trying to deliver to customers – all flies in the face of the accepted way of doing things in the homebuilding industry.

"Talk about coelacanth."

"Nice rant", said the CEO.

The intrepid, results-based consultant held on to the amused expression for a few moments, turned serious, and continued.

"Let me state this as clearly as I possibly can:  *The gains in productivity and – by extension – the creation of a sustainable degree of competitive separation that is the result of fixing the scheduling problems inherent in traditional project management methodology dwarf the gains that would result from any other improvement effort.*"

Traditional (Critical Path) PM v. Lean/TOC (Critical Chain) PM
<u>Single project</u> v. <u>multi-project</u>
Task Duration (reducing) and Hidden Safety (removing)
Buffers

"Let's talk about how we should schedule job/projects in a project portfolio – that is to say – in a multi-project environment", she said.

"If we were only concerned with scheduling a single job/project, the critical chain would be the limitation – the constraint, if you prefer – on how fast the job/project can be completed, because the job/project cannot finish any faster than the schedule on those tasks.  And, that would be the extent of our concern.

"However, homebuilding is multi-project in nature, which means that, at any point in time, the system – the portfolio of projects – has different projects at different stages, all competing for finite resources;  *that* is the concern.

"For example, our three superintendents in the room have a total of five houses at framing stage, but the framing contractor they use only has three framing crews.  Even if we initially schedule the jobs with consideration of the conflicts, any variation throws off the schedules of all the affected jobs.

"Since every job has a critical chain, and they all share resources, the critical chain of a single project cannot be the constraint for an entire system comprised of projects – the critical chain of a single project cannot be the constraint for an entire portfolio of projects.

"Instead, the constraint in a multi-project environment like homebuilding is the shared resource that has the least capacity or is in the highest demand, measured by the demand-to-capacity ratio.

"When we schedule jobs, we are scheduling the ordered sequence of tasks in multiple projects, based on dependency, defined by the predecessor-successor relationships.  The longest chain of tasks through the task network still determines the duration of each job, but the limitation on the rate at which RB Builders can generate Throughput – you recall the definition of that term from previous sessions, two sides of the same coin, etc – is determined by the capacity of the most constrained resource.

"We call that resource the *Capacity Constraint Resource* – the CCR, for short."

"Superintendents – they're resources, right?", asked the VP of Construction.  "We have always debated the job load a superintendent should carry – how many is too many, what's the ideal number, and so on.  At some point, does the superintendent become the CCR?"

"The CCR could conceivably be an internal resource, like a superintendent", said the intrepid, results-based consultant. "But, the Capacity Constraint Resource is almost always an external resource, like the framing sub-contractor we always seem to use as an example. We pick on him, but conceivably, it could be any sub-contractor. Since RB Builders plans and manages production at the community level, the CCR could even vary by community.

"Since homebuilding is more concerned with managing multiple jobs/projects in a portfolio, not single jobs/projects, it stands to reason that it is the condition, the capability and capacity, of the Capacity Constraint Resource that must sequence the release – the staggering, the pacing, the pipelining – of new jobs into the system.

"Remember – we have already described RB Builders' production system as a pipeline with two control mechanisms, with two valves. The first valve is the rate of closings, which authorizes each start and controls the level of work-in-process. The second valve is the Capacity Constraint Resource, which controls the timing of the actual release of new starts. So – not only is the Capacity Constraint Resource the constraint, it is also the pacemaker."

"I like this approach", said the senior superintendent. "I like it a lot. But, it raises some questions. First – how do we identify the CCR? Second – how do we keep it from moving around? We would have to schedule the release of new jobs on a continuous basis, so whatever resource becomes the CCR has to become more-or-less permanent. Correct?"

"The Capacity Constraint Resource is the resource with the least capacity and/or the highest demand", said the VP of Construction. "So – we just have to figure out which resource fits that bill. However, it's also an opportunity for us to select that resource, and plan and manage our production around it.

"If we want to know where the CCR is right now, it should be obvious – look at where the jobs seem to pile up, where they seem to queue", he said. "Or, just ask ourselves which resource we tend to fight over the most. My point is, if we choose which resource we want to be the CCR, then we will know where it is, and I don't think it will tend to move around.

"Variation and uncertainty, as we know, are a fact of life. We eliminate as much of it as we can, and we manage the rest. Choosing the identity and location of the Capacity Constraint Resource makes it easier to plan and manage production."

"Agreed", said the senior superintendent.

*Traditional (Critical Path) PM v. Lean/TOC (Critical Chain) PM*
*Single project v. multi-project*
*Task Duration (reducing) and Hidden Safety (removing)*
*Buffers*

"Let's talk about the how the duration of tasks should be determined", said the intrepid, results-based consultant. "We've seen the problems that occur with task durations based on supposedly high levels of certainty, seen how padded task durations intended to protect the completion date actually results in longer-than-necessary durations and project schedules, yet don't necessarily result in projects finishing on-time.

"We need to reduce task durations and remove hidden safety.

"The idea is to not schedule tasks with 95% confidence", she continued, "Doing so spreads too much safety throughout each job.  For naught, because it doesn't provide a means for managing variation.  The intended safety is wasted.  So – not only do jobs take too long to complete – they also finish late.

"Remember your framing sub-contractor?  He's thinking 10 days, but tells us 20 days, where he is 'very confident', 95% sure.  Statistically, he has a 50/50 shot at completing the work in about 12 days.  So, removing the safety – represented by the difference between a 50% confidence level and a 95% confidence level – reduces the duration of the task;  in this case, by about eight days.

"However, targeting a 50% probability of completing a task or job on schedule – on time – is not an acceptable outcome.  There has to be more stability to a process than only being able to complete a job, on schedule, half the time.

"What we need is a way to have our cake and eat it, too;  yin and yang;  the power of complementary opposites.  What we need is a way to achieve the shorter duration of a job schedule with no padded individual task duration, with the 95% confidence of a job schedule if it did have padded individual task durations.

"The good news about a 50% probability is that it is, in fact, a 50/50 proposition;  half of the tasks can be expected finish on time, or even earlier.  Earlier finishes would offset late finishes.  So, we don't really need more safety in each task.  What we need is a way to protect the due date of the overall job schedule from the variation that would cause it to finish later."

*Traditional (Critical Path) PM v. Lean/TOC (Critical Chain) PM*
*Single project v. multi-project*
*Task Duration (reducing) and Hidden Safety (removing)*
*Buffers*

"Keep that thought in mind, and let's talk about buffers", said the intrepid, results-based consultant.  "Buffers are the devices that enable us to protect a system from variation and uncertainty.

"We have previously discussed the 'physics' that demonstrate how a production system will protect/buffer itself from variation and uncertainty, through some combination of additional inventory, excess/reserve capacity, or longer duration.  It is common to every production system, to every approach to production thinking.  It's part of their DNA.  It is known as the Law of Variability Buffering, and the tendency to pad the individual task durations of every job schedule is a classic example of that law of production physics.

"Rather than try to protect every task of a project, or every phase of a job, from variation and uncertainty with a hidden safety buffer of padded duration, it would be better to reduce task durations to 50% probability (the percentage at which there is no safety), and then take a portion of that resulting shorter duration and place it where it can be used.

"It is the same theory of pooled risk that insurance companies use."

"Are we going somewhere with this?", asked the VP of Sales.

"Of course", replied the intrepid, results-based consultant. "Instead of requesting, expecting, and accepting task duration estimates with 95% probabilities, what if we asked sub-contractors to provide task duration estimates under a set of positive assumptions that would produce 50% probabilities?"

"The frightening prospect of promoting 50/50 probabilities notwithstanding, 'positive assumptions' are always a good idea", said a superintendent, sarcastically.

"You mean, of course, could I be more alliterative?", she asked, patiently. "Most people find it difficult to think in terms of probability of outcome when estimating duration. Instead, they need to be asked for task duration estimates based on 'positive presumptions'. There, is that better?

"Tell them to base their task durations on presumptions like these."

She made a list.

*Have everything you need.*
*No multi-tasking – focus only on the task at hand.*
*No surprises.*

"Here's what you tell them: First, give me your 'very certain' durations, durations in which you can finish your work on schedule 95%+ of the time. Then, give me your task duration estimates, under these three conditions: (1) presume you will have all the information and material you need to do the job, (2) presume you are able to focus on this task only, without interruption, and (3) presume there will be no surprises that add work or lengthen time.

"Using this set of presumptions will usually produce an estimated task duration with something close to a 50% probability, without raising the specter of it."

She continued. "What if we then told these sub-contractors that we realize there is no safety in their 50/50 probability task estimates, and that we realize and expect those estimates to sometimes result in – statistically, about half the time – completion times that exceed the scheduled duration for their tasks?

"And – what if we assured them that it was alright to give us those types of estimates? What if we said we expected – we, in fact, required – 50% probabilities?

"Then, what if we took a portion of the difference between the estimates with 50% probability and the estimates with 95% probability – let's say, half of it – and put it in a buffer to protect the overall scheduled completion date of the project/job, as opposed to futilely trying to protect the completion date for each individual task with all of the padded durations?"

The intrepid, results-based consultant reached across her notebook computer and recovered her programmable HP-12C – the most recent version of the financial calculator she had long carried with her as a weapon – from the superintendent who had been fiddling with it.

"Figure it out?"

"It's an HP-12C. It uses RPN. It's been around for more than 25 years, almost as long as me", said the superintendent. "It's old school. It might be a coelacanth."

"Definitely old school, but it's not a coelacanth", she said. "That's because it still works better than anything else that has come along. Unlike the algorithms still in use by most scheduling programs. There is a difference between a classic and a fossil.

"But – to think that I thought you guys only counted on your fingers and toes. So, where were we? Ahh, yes. Remember the product we were talking about earlier?

"As I recall, RB Builders takes close to 180 days to complete a house, despite a job schedule that is intended to be only 120 days. Since 120 days reflects the sum of highly-confident task durations, it stands to reason that about half the time, those jobs could be completed in around 74 days. As you remember, we characterized the 180 day actual cycle time as a picture of all the delays cascading through the overall job schedule.

"What if we reduced the job schedule from 120 days to 74 days – a difference of 46 days – and then extended the new, shorter job schedule to 97 days, by inserting a 23 day project buffer at the end of the job schedule?

"What would happen?"

"It would save 21 days out of the schedule", said the superintendent. "Didn't need a calculator to figure that one."

The CFO looked at the intrepid, results-based consultant. "Our actual cycle time is 180 days, not the 120 days called for by the job schedule. Are we to draw from this that taking half the padded duration out would reduce our actual building time from 180 days to 157 days?"

"I don't think that's what she's saying", said the senior superintendent.

"What she's *saying*", said the VP of Construction, "is that we can reduce our actual cycle time by an average of 83 days, from 180 days down to 97 days, and in the process, make the schedules far more reliable, by cutting out a lot of the hidden safety that the schedule consumes anyway, because of the human behavioral tendencies we talked about before. Remember, the only reason we experience 180 day cycle times on jobs with 120 day schedules is variation and uncertainty; eliminate the negative effect of variation and uncertainty, and you can do exactly what she is saying."

"Save 21 days on the schedule, cut 83 days out of the actual build time", said the senior superintendent. "Not too shabby."

The intrepid, results-based consultant smiled.

"That's right", she said. "Without doing anything else differently – with the possible exception of making all of your jobs easier – you would cut your cycle time almost in half on every house you build. Without adding one iota of production capacity – from either a resource standpoint or a financial standpoint – productivity would almost double. Which means, the rate of physical throughput – the rate of closings – would likewise almost double.

"In today's real estate market, we don't know what that means for Gross Income or Net Income, because we don't know what kind of margins it would take to acquire all of the sales necessary to fully utilize the additional capacity that would be the result of that much increase in productivity.

"What we do know is that – as long as the Gross Income Margins are positive – all of the Gross Income above breakeven would drop straight to RB Builders' bottom-line.

"Not too shabby."

"Given the current market conditions, we clearly have a situation with excess capacity that we probably won't be able to reconcile", said the CFO. "This is definitely not the smart short-term play. We should be bleeding-off excess capacity and overhead cost, not finding ways to become more productive and create even more unusable capacity.

"I do agree with the point you made earlier. Long-term, the gains in productivity that result from fixing the scheduling problems inherent in traditional project management have the potential to far outstrip the gains that might result from anything else we can do to improve operating performance.

"I'm worried about now."

"We know that", said the CEO. "This is about securing the long-term future of the company. We are working on other rapid-results projects with much faster paybacks.

"This is about creating sustainable competitive separation into the future."

"I'm not buying it", said another superintendent. "180 day cycle time is a fact. 120 day job schedule is also a fact. However, I don't see how the 23 days that comes from reducing the duration of tasks from 'highly certain' to 'toss-up' could possibly reduce overall duration by 83 days."

"Let me explain it to you, once again, buckaroo", said the intrepid, results-based consultant. "It's due to a combination of things. The padded duration that we hide in each task is unmanageable. The tendencies of human behavior conspire to use it all up. In a system, delays multiply and cascade, due to all of the interdependency.

"The traditional CPM-based method only accounts for task dependency, and has no way of dealing with resource conflict, other than to put float in the schedule; CPM says, 'makes sure you do A before B, but if you have A doing three things at once, we can't help you'. If the templates had float tasks inserted, the float would represent more time, which the tendencies of human behavior still conspire to use up.

"The problems with systems are much more than the sum of the problems of the individual parts.

"The mere fact that everyone in this room acknowledges that RB Builders should be able to build a home in 90 days, its schedules stipulate 120 days, but it takes the company 180 days should tell you this is true. If the float was actually in the schedule as separate tasks, the schedule would go to somewhere between 120 and 180 days, say 150 days, and I guarantee you the actual cycle time would increase to 220 days, because nothing has changed."

"According to you guys, I'm just a glorified sales representative", said the VP of Sales. "That's what I used to be, and maybe some of you just think that I have risen to my level of incompetence.

"Personally, I think of myself as a rainmaker."

"Rainmaker – now, that's a good one", said the CFO. "We're in the middle of a drought, Chief. You need to break out the rain dance."

"The point is, I'm not supposed to think in these terms, because it's not my area, supposedly not my forte'", said the VP of Sales. "If these numbers are right, then someone needs to explain why everyone on the production side knowingly tolerated cycle times that were twice as long as they should be."

The VP of Construction shrugged his shoulders, smiled, and allowed, "I cannot enlighten you, Rainmaker. There is no acceptable answer. I must but presume that the Age of Intrepidness had not yet come upon us. Perhaps, the duration devil will be found in the details."

"Since the Age of Intrepidness seems to have finally arrived, I'm sure our sage consultant could tell us how the job/project schedules have to be changed to achieve shorter durations", said the superintendent.

The intrepid, results-based consultant restrained herself from making any response or comment, and simply made a numbered list.

2. Durations with 50/50 probabilities – no task safety
3. Require ALAP task starts
4. Eliminate ASAP task starts

5. Eliminate task due dates
6. Concentrate safety in project buffers
7. Eliminate Parkinson's, Student Syndrome, and Multi-tasking
8. Deal w/ resource contention
9. Manage buffer penetration

"You missed one", said the superintendent.

"The man has a mind like a steel trap", she said. "The numbering starts at '2', because number '1' on the list is to make RB Builders' existing scheduling software compatible with Critical Chain Project Management.

"Numbers '2' through '9' are all actions that occur afterward."

Looking at the VP of Construction, she continued. "You mentioned two problems regarding the existing scheduling application in the earlier discussion. The first problem dealt with buffers, or more precisely, the absence of buffers. You had to insert float tasks, because the software had no means of inserting buffers. The templates didn't have built-in float tasks, either. Not having built-in float tasks is not really a problem, since we don't want to use float. But, float doesn't do the job of protecting the completion date. Float only adds duration.

"The second problem was about the algorithms, which use critical path, not critical chain. Now, *that's* a problem, because the critical path method has deficiencies – deficiencies that the critical chain method overcomes.

"As I explained earlier, there are ways to make critical path-based programs perform critical chain scheduling. Unless you choose to replace your existing scheduling application with a standalone critical chain application, making your existing application perform critical chain scheduling will require an add-in application.

"Whether you elect to use a standalone application or an add-in application, the software conversion has to be done before we can do much of anything else. The rest of it needs to be done in the order listed. These steps describe what you have to do to change your job/project scheduling from CPM to CCM, but, first, you have to have software that can perform critical chain."

The intrepid, results-based consultant drew the team's attention back to the first list on the erasable board. "In many ways, the area of job/project scheduling is the crux of project management", she said. "I can tell you that it takes a lot of work to get it right and become proficient at it.

"Let's switch the subject to how you manage projects." She underlined the last item on the erasable board.

Project Duration
Project Structure

Project Scheduling
Project Management

"Project scheduling and project management are obviously connected, since one follows the other", she said. "The difference between scheduling projects and managing projects is a function of the objects on which they focus. In project scheduling, we create a job and schedule the tasks required to complete that job. *Project scheduling* includes setting task durations, setting task dependencies and relationships, resolving any resource contention, and setting up buffers to protect the job completion date. *Project management*, on the other hand, is about monitoring and taking action on certain things.

"We'll get to what needs to be monitored and acted upon in a minute. But, before we do, I want to briefly go back to some areas we talked about in previous sessions.

"Both job scheduling and buffer management occur in the context of a multiple-project/project portfolio management environment. Because, that's what homebuilding is – managing multiple jobs. Not one job, not multiple jobs independent of each other. It's about managing a portfolio.

"Remember the discussion on pipelines and production systems?"

She turned around and rummaged through the discarded flipcharts stacked behind the easel containing the lists that the CEO's administrative assistant had been dutifully transcribing from the erasable board at the conclusion of each session.

At last, she found the list she wanted.

Size = Work-in-Process
Length = Cycle Time
Capacity = Closings with a controlled level of WIP
Control = Rate of Closings and capacity of the scheduling resource
Cost = Operating Expense and Resources

"There is a pipeline, and that pipeline has a determinable size, capacity, length, and cost", she continued. "We said that the *size* of the pipeline is measured in terms of the amount of work-in-process that it carries. Its *capacity* is measured by the rate of throughput – the output – it produces. Its *length* is measured in terms of the number of days from start-to-completion – what we call cycle time, or, what we sometimes call duration.

"The *cost* of the pipeline is determined by all the cost of owning it, represented by all of the indirect, non-variable costs we commonly call overhead.

"There are the two control mechanisms – or *valves* – that pace, or stagger, the release of work into the pipe, in order to make certain that the pipe does not exceed its size or its length, yet allows it to utilize its capacity and its cost. The first valve is the rate of closings; the second valve is the capacity of the scheduling resource, which we call the CCR.

"So, let me ask you boys a question", said the intrepid, results-based consultant, directing it to the superintendents and VP of Construction. "As it now stands, what is it that you would say you actually 'manage' in the RB Builders' production system?"

"We manage our jobs", replied a superintendent. "We manage the jobsite. We manage the job schedule. We manage the job quality. We manage the job budgets."

"Fair enough", she said. "That was too broad. Let me rephrase it. Among your other job-related tasks, you say you manage the job schedule. What is it about the job schedule that you currently manage?"

"We manage the tasks", replied the superintendent. "We manage the tasks according to the job schedule, which means we get the tasks started as soon as we can, and we try to get the tasks finished on-time. Somewhere in all of that, the resource conflicts you were describing get worked out, not always to everyone's satisfaction."

"Since you manage the tasks on each job", she continued, "then I presume you also know the completion date of each task, the start date of the next task, the completion date of the job, and whether or not the job will finish on-time.
"Am I right?"

"I'm sure we think we do, but we really don't know completion dates, not with any level of confidence", said the VP of Construction, glowering at the cluster of superintendents. "She's right, and you guys know it. We have already admitted that we don't come close to meeting our completion dates, even though we all agree that our job schedules provide significantly more time than we need to build a house."

"The problem is, every time a task finishes late or is delayed, the job schedule moves the job completion date out. The revisions to the job schedules are constant. But, we still don't know whether the job is going to be completed within even the new time. There is variation and uncertainty in every job, and it extends all the way through it.

"Regardless of whether the first 10 tasks finish early, on-time, or late, there is no way to see where we really are in terms of overall completion, because the next 10 tasks could be a completely different outcome. The padded duration of each individual task only protects that task – not very well, I might add – and does nothing to protect or insure the on-time performance of the job/project.

"Clearly, there are problems. We know that.

"Bottom-line, with the current project management methodology, there's really no way to manage projects in a project portfolio", he said. "I'm sure you're going to tell us that it's about doing all of the stuff on your numbered list – 50/50 task duration estimates, late-as-possible starts, buffers, etc. – but it sounds different, and it seems complicated.

"It's all quite disconcerting."

"Buffer management", said the intrepid, results-based consultant.
"What do you mean?", asked a superintendent.

"Managing projects in Critical Chain Project Management is about one thing, one thing only", said the intrepid, results-based consultant. "Buffer management. It's about how you manage penetration of the buffers that protect various elements of the job schedule, and making decisions based on how much safety has been consumed – how much buffer has been penetrated – by delays.

"You manage the buffer, not the schedule.

"If you recall, under the new approach to scheduling projects – rather than try to protect every task from variation and uncertainty with 'highly-probable' durations, each containing its own built-in, hidden safety, in the form of padded duration or float tasks – you are going to reduce task durations to 50% probability, give up the task safety, and then take a portion of the sum of difference between the 95% confidence and 50% confidence levels, and put it into a buffer where it can actually be used.

"Buffers have been explained from the perspective of how a production process protects itself from variation and uncertainty, with a combination of additional inventory, excess/reserve capacity, and longer duration. Those are *categories* – means – of protection; we also need to talk about the *types* of buffers, which partly has to do with location and action.

"In Critical Chain Project Management (CCPM), there are four types of buffers, three that are common to both single and multi-project management, and one that applies only to multi-project management", she said, making a list.

Project Buffer
Feeding Buffer
Resource Buffer
CCR Buffer

"The first type of common buffer is the *project buffer*, comprised of the additional duration added to the end of the task network to protect the completion date of the overall job; it is the most important buffer we have to manage. The project buffer is created from a portion of the time/duration removed from all the individual tasks.

"The second type of common buffer are *feeding buffers*.

"Wherever a job has a set of tasks that run concurrent/parallel and then merge back into the critical chain of tasks that determine its overall length, a buffer is inserted to make certain that one feeding chain cannot delay the entire job/project.

"The feeding buffers are time buffers; they do not lengthen the job schedule, because they are not on the critical chain. They do, however, prevent the behaviors that make projects take too long and finish late; they force ASLAP starts and they prevent multitasking.

"The feeding buffers force resources to focus on doing the work that has to get completed, in order for the job not to be delayed.

"The third type of common buffer are *resource buffers*.

"A resource buffer is about notification, kind of a 'heads-up, we're going to need you in three days' type of communication; basically, it's an alert. Resource buffers do not have a time element, do not affect duration.

"There is one other buffer, which is called the *Capacity Constraint Resource buffer*. It's an important buffer, also.

"The presence of the Capacity Constraint Resource (CCR) was explained earlier. In a multi-project environment – where we have a portfolio of projects that we have to manage – the CCR is what determines the throughput of the production system, because it is the system's most constrained resource[9]. The CCR is the system's constraint resource because it either has the least capacity or the highest demand of any resource in the system.

"More importantly, the CCR is also the scheduling resource, the resource that must sequence the release – the pipelining – of new jobs into the system. So, while the CCR is both a constraint and a pacemaker, its responsibility is to function as a drum/pacemaker.

"The situation we need to avoid is overloading this resource, by releasing too many jobs into the production system, and then having them queue at what becomes a bottleneck.

"The Capacity Constraint Resource buffer is the mechanism that protects the CCR from the variation and uncertainty that would have that effect", said the intrepid, results-based consultant. "In many project management organizations, the Capacity Constraint Resource buffer would consist of a block of time somewhere between the use of the CCR on one project and the scheduled use of the CCR on the next project; those blocks of time – those CCR buffers – would not lengthen the duration of any of the projects in the portfolio.

"RB Builders does not need a time-related buffer; we need a mechanism that releases new jobs into the production system at a pace that does not overwhelm our most constrained internal resource. For our purposes, the CCR buffer is a buffer that considers reserve capacity, not time.

"Creating this mechanism is saying: 'This is the most constrained internal resource in the system; this where it is located, in this job phase, it is this sub-contractor. It has a planned, demonstrated, contracted capacity, and we are going to load it – schedule its work – at a rate that should not exceed that capacity, in order to protect it from variation'.

"For the record, scheduling a production system at 100% capacity makes it very unstable, particularly if that system tries to insure 'balanced capacity'; statistically-speaking, instability begins to occur anytime the utilization rate rises above about 75%.

---

[9] The Capacity Constraint Resource is obviously an internal constraint. If the system constraint is an external (market) constraint, the CCR is still the internal resource with the least capacity or the highest demand, but it would not be the system constraint.

"Therefore, if we do not give the Capacity Constraint Resource some protection – i.e., a mechanism that insures that we don't start jobs at a rate that exceeds the capacity of the constraint – then queuing theory assures us that work-in-process will pile up behind it, which makes for a bigger company and lengthens cycle time, two undesirable effects that we clearly do not want to occur.

"A time-related buffer would solve that problem indirectly, but a Capacity Constraint Resource buffer that considers reserve capacity – i.e., a start rate of new jobs that does not exceed the capacity of the constraint – is a better approach."

"This is kind of new for all of us, so I want to make sure we all understand it", said the senior superintendent. "In our visual image of the pipeline, the CCR is the pipe's second valve. Its purpose is to control the timing of the release of new starts previously authorized by the rate of closings. Although it is atypical, the staggering of starts according to the pace of production (determined by whatever resource is our CCR) is a better approach to dealing with variation than a time-related buffer, in terms of optimizing productivity, maximizing throughput, controlling work-in-process, and reducing cycle time.

"We manage – we make decisions – based on the status of the project buffer, not the individual job schedules.

"Correct?"

"Yes", the intrepid, results-based consultant said. "Managing the CCR does involve both pace and protection. You almost can't talk about pace and protection independently; they are sort of joined at the hip. The protection is the proper release pace.

"However, I do need to get this discussion back to buffer management.

"Buffers are created when schedules are initially built, part of the job schedule template, part of each job schedule. Managing multiple jobs/projects is a function of monitoring and acting upon the dynamic changes to the project buffer, as the schedule intrudes on it. This intrusion is known as buffer penetration, and the consumption of the safety that results from the intrusion is what has to be managed.

"Buffer management simplifies the task of keeping jobs/projects on schedule; it requires accurate information on the amount of work remaining on each job, not the percentage of work completed. To that end, the project buffer requires visual management tools, so that you can 'see' the ratio of work remaining to buffer remaining, reflected in the progress-adjusted buffer; and, by progress-adjusted, I mean the total buffer scaled by the proportion of work remaining."

She pulled up an Excel chart on her notebook and projected it on the conference room screen.

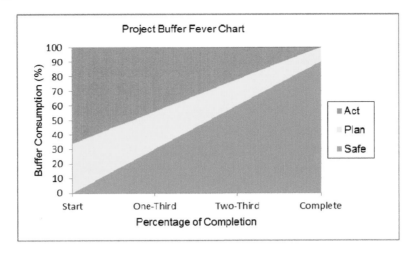

"That tool is called a *project buffer fever chart*, buffer usage on the y-axis, job progress on the x-axis", the intrepid, results-based consultant explained. "The fever chart is a dynamic control chart, showing whether you are within the tolerance on consumption.

"It's kind of a green-yellow-red thing."

"You mean, like the flag of Bolivia", said the VP of Sales. "Or, the flags of half the provinces of Ecuador."

"Like a traffic light", said one of the sales representatives.

"I think you have the picture", she said.

"A fever chart is a gauge. If a job is, let's say, 85% complete and has 50% of the buffer remaining, that job is likely going to finish on time. It's in the green. Focus the effort elsewhere. On the other hand, if that same job is 50% complete, but it has consumed 85% of its buffer, it needs to be expedited. It's in the yellow. Focus the effort there.

"As a rule of thumb, if usage exceeds one-third, 33%, of the progress-adjusted buffer, someone better be figuring out what to do about it, how to recover it. If the usage grows to two-thirds of the progress-adjusted buffer, the project falls into the red, and the recovery plan needs to implemented."

"The 'progress-adjusted buffer' means what, again?", asked the VP of Construction.

"You scale the buffer requirement as the job progresses", she replied. "The less work remaining, the less aggregate protection you need. Let's just look at the situations that would exist on a representative selection of your jobs", she continued, quickly sketching a data table.

"Let's say we are doing critical chain scheduling, and we've already recast the job schedules to reflect 50/50 task estimates, created the project buffer, etc. Instead of the old 120 day schedules, all of these jobs now have the new 97 day completion schedules we discussed earlier, including a 23 day project buffer. In order to make the percentages easier to calculate, let's just say the jobs in this illustration have 100 day schedules, including a 24 day project buffer.

"We'll use the hypothetical contract owners' last names to identify the jobs."

| Job | % used | Days left | Buffer used | % | Condition |
|---|---|---|---|---|---|
| Smith | 80% | 20 | 0 | 0% | safe |
| Jones | 60% | 45 | 5 | 45% | caution |
| Thomas | 50% | 45 | -5 | 0% | safe |
| Bradley | 40% | 72 | 12 | 68% | warning |

"So we're clear, the term '% used' means the percentage of the job schedule that has been consumed, although not necessarily achieved. The term 'days left' refers to the amount of work remaining, measured in days. When it says 'buffer used', that is the calculation, in days, of progress-adjusted buffer penetration/consumption, while the next column, marked '%', denotes the percentage of the progress-adjusted buffer consumed.

"The term 'condition' is a function of buffer penetration, adjusted to the actual job progress – if it grows to one-third, you have to watch it, if it grows to two-thirds, you need to take action.

"So – what are we looking at with these four jobs?"

One of the superintendents walked to the board.

"Smith has used 80% of its schedule and is also 80% complete, which means that there are 20 days of work remaining", he said. "It has not used any of its progress-adjusted project buffer. Smith is on schedule. Therefore – if I understand the 'buffer-adjusted-for-progress' mentality and have calculated it correctly – the buffer has not been penetrated.

"So, Smith is safe."

"What about Jones?", she asked.

"Jones has used 60% of the job schedule, but has only completed 55% of the work", said the superintendent. "Or – we could say that Jones has 45% of the work remaining, but only 40% of the schedule remaining in which to complete it. In any event, Jones has consumed five days of project buffer. Adjusted for progress, Jones' project buffer should have about 11 days remaining, but it only has six days. The buffer penetration is about 45%, which means that we should already have a plan in place to recover the buffer and get the job to finish on schedule. Jones is in a bit of trouble."

"Thomas?"

"Obviously, one of my jobs", he said, with satisfaction. "I've only used 50% of my scheduled days, but I have completed over 55% of the work. So – I have a five-day buffer surplus. It'll finish well ahead of schedule.

"Thomas is in great shape."

"And, what about Bradley, who appears to be the big winner in the 'Penetrate the Buffer' sweepstakes?", asked the intrepid, results-based consultant.

"40 days into the job, 60 days of schedule left, and they have 72 days of work remaining", said the superintendent. "12 days behind, and they aren't even halfway yet. That's – what? 68% of the project buffer, adjusted for real job progress?

"Someone better throw the switch on the buffer recovery plan for Bradley."

"What kind of alarm are we talking about here?", asked another superintendent. "It better be cowbell. 'Cause I gotta FEVER CHART! And, the only cure is more COWBELL!!"

"Don't you have anything better to do on weekends?", asked one of the sales representatives.

"I gotta question", said the intrepid, results-based consultant. "Since both Jones and Thomas are now at exactly the same stage, how are you going to resolve the resource conflict that situation will inevitably cause?

"More cowbell?"

"What conflict?", asked the superintendent. "Our values don't allow for conflict to occur; it's against the rules. Both jobs would be scheduled for ceramic tile at the same time, so I presume that's the conflict we need to resolve. Jones started before Thomas, so Jones gets the ceramic tile installer first.

"That's even-flow production – keep all the jobs evenly spaced, don't let one get in front of the other. That's FIFO; it's a build-to-order process."

"Right answer, wrong reason", said the VP of Construction. "We're going to send the ceramic tile installer to Jones, because it has the biggest project buffer penetration, not because of the start dates. We want to maximize the rate of throughput without increasing capacity or inventory – 'more-for-less'. The only way to accomplish that is to make jobs finish on schedule. Only when the ceramic tile installer is finished with Jones, can he go to Thomas."

"This little exchange does bring up a common misconception", said the intrepid, results-based consultant. "You mention even-flow production, and plenty of folks think it's this rigid, evenly-spaced system, like everything fits into slots. I recall the description of RB Builders' previous production system. It was characterized as even-flow production, but it was achieved with a start slot matrix. The start matrix created a push system. It didn't result in even-flow.

"Remember – even-flow is an outcome, not a mechanism.

"When I hear answers that adhere to a rigid sequence – A-B-C-D – with even spacing, I have this picture of a freight train with all of the boxcars coupled to each other, every car moving at the same pace, each car locked into its designated place.

"When one car moves, they all move. When one car stops, they all stop.

"This so-called 'boxcars' method is one of the earliest versions of even-flow production in the homebuilding industry, but it is even-flow production that pays an enormous price, in terms of productivity and output.  It is even, but it is very long and very slow.  It imposes requirements to community-based production planning and management that is very rigid and inflexible."

The intrepid, results-based consultant returned to her numbered list, underlining the last two points.

2. Durations with 50/50 probabilities – no task safety
3. Require ALAP task starts
4. Eliminate ASAP task starts
5. Eliminate task due dates
6. Concentrate safety in project buffers
7. Eliminate Parkinson's, Student Syndrome, and Multi-tasking
8. Deal w/ resource contention
9. Manage buffer penetration

"So that we are clear", she said.  "Resource contention – which is our fancy term for when a resource is required to be two places at once – is anticipated and addressed when the schedule for an individual job/project is created.  The priorities of the resources are managed – allocated – daily, in relation to the conditions of the project buffer – how much of the buffer has been consumed, how deeply the schedule has penetrated the buffer.

"Also – we manage the condition of the buffers, *not* individual job schedules or the individual tasks within them", she continued.  "Once established, an individual job schedule is not adjusted, until or unless the project buffer is completely consumed.

"Managing buffers, as opposed to managing schedules and tasks, is a major departure from traditional project/job scheduling, where the schedule itself is constantly adjusted, as tasks finish late.  Eliminating the need to constantly revise job schedules based on start dates and completion dates makes the process much more stable.

"It significantly reduces the amount of variation and uncertainty, because it takes the moving parts out of the job schedule."

The intrepid, results-based consultant added an item to the numbered list.

10. Relay Race

"Throughout these sessions, we have been referencing and comparing two production management approaches, both of which have application in homebuilding", she said.  "RB Builders' production system is a blend of these two approaches, crafted to meet the specific requirements and parameters in which the company must operate.

"Lean Production and Theory of Constraints both make use of analogies to depict the thinking behind their approaches to managing production, from both a process and project perspective. There is particular agreement – Taiichi Ohno's 'relay racer', on the Lean/TPS side comes to mind – regarding the analogy that describes a relay race. As with almost anything else, analogies comparing production to running a relay race can be taken too far, but the relay race analogy is particularly useful for project management.

"In a relay race, the goal of the team is to win the race, not run to meet a specific time for each leg", she explained. "The race has a start, and then every teammate runs their own leg as fast as they can.

"After the race starts, none of the runners pay any attention to the clock, nor do they pay any attention to specific start times or finish times. They certainly don't participate in other events while they are running the race. They prepare to run – they get themselves in the handoff zone and get ready for the exchange – and, once the exchange is made, they simply run as fast as they can.

"For just a moment, try to envision what a relay race would look like under traditional project management.

"Traditional PM – captured in the critical path method – would give each runner a required time to post that each of them are 'very confident' – are 95% certain – of achieving. But, knowing they have time to spare, natural human behavior would kick-in. Runners would be inclined to postpone starting – Student Syndrome. They would be inclined to run just fast enough to meet the required time – Parkinson's Syndrome. They would be inclined to do something else in addition to merely running the race – multi-task.

"As a result, the best each runner would likely achieve is to meet their required individual time, the aggregate of which would be too slow to win the race", she continued. "More than likely, some of them would fail to meet their individual required times, and they would have an even slower overall time.

"And, in the rare occurrence of an earlier-than-required finish on one leg of the race, the next runner would probably not be ready, willing, or able to take advantage of the early finish.

"Now, compare that approach with how the Lean/CCPM approach would handle a relay race.

"Under Lean/CCPM, when a runner reaches the handoff zone, the two runners synchronize the baton exchange, losing as little speed as possible, transitioning as fast, as smoothly, as seamlessly, as possible. When the first runner is nearing the end of his leg, the next runner has to know to be in the zone, ready for the baton exchange, ready to run. Since each runner runs their leg as fast as they can and they seamlessly synchronize their exchanges, the team can capitalize on faster-than-previous times, and either post a faster overall time or, at least, offset a slower leg, and still win the race.

"Traditional PM, on the other hand, would have the 'successor' runner starting too soon or too late, but in either event, probably not in synch with the 'predecessor' on the exchange. As a result, the two runners – and, therefore, the team – would lose time on the exchange; the runners might miss the exchange altogether. The slow times would accumulate, but the faster times would be wasted.

"Under either the Lean or the Theory of Constraints approaches to scheduling projects, event conflicts would have been proactively resolved before the race even started. Contrast that with the CPM method of scheduling used in traditional project management, where resource contention is not considered, and the runners may not even show up!"

The intrepid, results-based consultant set the erasable marker down, closed the project file, and started to shutdown her notebook. When she was finished, she stood and surveyed the room. She noted, with some satisfaction, that every available inch of wall space was covered with flipchart paper, and that every inch of erasable board was full.
Then, she spoke, as she highlighted each area.

"We talked about the differences between project management and process management, particularly in how those areas of workflow are scheduled", she said. "We talked about Critical Path and Critical Chain, the differences. We talked about the elements of human behavior that affect production systems – padded durations, Student Syndrome, multi-tasking, Parkinson's Law, how Murphy enters the picture.

"We delved into the elements of projects – duration, structure, scheduling, and management.

"We talked about buffers, the steps in developing Critical Chain schedules. We went back-over the visual image of a production system as a pipeline, the elements of that pipeline. We talked about the different types of buffers, and their purposes. We talked about the analogy of the relay race.

"We talked about cowbell.

"That was a lot to cover in one session, but that's it for now", said the intrepid, results-based consultant. "We have one final session that will help you see the principles of process management and project portfolio management more clearly, and pull everything together for you.

"As always, you did a great job."

# CHAPTER VII: "THE GAME"

The intrepid, results-based consultant opened the portfolio, removed a set of game boards, and leaned them against the conference room wall. Rummaging through her notebook carryall, she located and removed the zippered pouches she would need. Two of the pouches contained poker chips, and the third contained multiple, different sets of six-sided dice.

"I need the conference room table cleared", she said.

"Change is a necessary condition to any improvement effort. You can't expect to get different results by doing the same things the same way. But, change is difficult, disruptive, time-consuming, and costly; plus, the effort can fail to produce the intended result.

"What we need to do is to be able to learn what to change, what to change to – excuse the dangling preposition – and how to make the change; in essence, we need to be able to discover, to figure things out, without the cost, disruption, and risk of failure associated with doing it in real life. What we need to do, is significantly reduce the learning curve on change.

"At the end of the previous session, I told you that I would help you pull all of the production principles and disciplines together in a way that you can use to more effectively manage production", she continued. "The new approach that you will use to manage production – and to thereby improve operating and financial performance – becomes intuitive and simple in practice, but there is a lot to understand.

"It is an approach that is counter to what most of you have been taught; initially, it can be difficult to grasp. In short – it must be learned. And, that kind of learning is a harsh teacher when it occurs at the cost of real operating performance and actual business outcomes.

"So – we are going to engage in some discovery-type learning, meaning the type of learning that occurs in business games[10] designed to simulate the competitive, fast-paced, rapidly-changing, uncertain, risk-laden, variation-filled environment – hope I didn't miss anything – in which production decisions must be made.

"It creates learning based on what you experience and do, not simply what you hear and read.

"Games can compress the learning curve into a span of days, by simulating production situations that speak to the circumstances you encounter in the real world – situations that are simplified in

---

[10] This game is adapted from the Project Dice Game, introduced in *Project Management in the Fast Lane: Applying the Theory of Constraints* (Robert C. Newbold, St. Lucie Press/APICS Series on Constraints Management, 1998), and elements of "The Synchronous Flow Game", a product of Chesapeake Consulting, Inc. (1999), distributed through APICS.

structure, fast to run, easy to see and understand, that we can modify and run again, over and over, until we get it figured it out, until we see and understand the principles."

She pulled up a PowerPoint slide on her notebook, projecting on the conference room screen an image of the first game board.

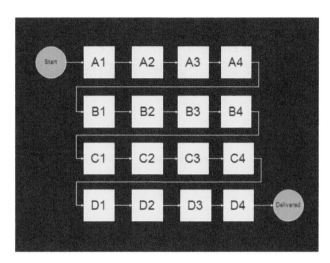

"We have here a very simple production system, one with balanced capacity", she said. "You can look at a production system as either a process or a project portfolio. We have talked about the differences in how processes and project portfolios are managed, and we noted during several of the previous sessions that the production aspect of homebuilding most closely resembles project portfolio management with embedded processes; although there are distinctions to be made, the production principles and disciplines we have been learning – the production physics – would apply similarly in either case.

"In this game, let's just say we are dealing with a production system that contains projects or jobs – that's what the system operates upon – and those projects/jobs result in completed houses."

She started a list on the erasable board.

*4 Resources (A, B, C, D) performing work in 4 Phases; balanced capacity*

*4 Tasks in each Phase – A1, A2, A3, A4 (complete), etc.*

*4 Resources X 4 Tasks = 16 Tasks for each Project/Job*

*Resources work only on Projects/Jobs completed by the previous Phase*

*Reverse Sequence (D-C-B-A)*

"This production system has four Resources – A, B, C, and D – each of which performs four discrete tasks in its phase of the work. In other words, Resource A has tasks A1, A2, A3, and A4 in its phase, Resource B has tasks B1, B2, B3, and B4 in its phase, and so on. One more time; four Resources, who each perform four tasks. So – four Resources times four tasks (in its phase of the work) equals a total of 16 tasks in each project/job. The completion of the A4, B4, C4, D4 tasks signifies that the respective Resources have completed all of the work in their phase for a particular project/job – they become the project milestones that we will call

'A Complete', 'B Complete', etc.  The completion of task D4 – 'D Complete' – also signifies the completion of the project/job – project delivered, all phases complete.

"Each Resource must complete all four of their phase's tasks in the proper sequence on a project before that project – that job, that house – can be available to the next Resource.

"Work is performed in the following order:  D-C-B-A.  Resource D does whatever it can, then Resource C gets to do whatever it can, then Resource B, then Resource A.  That means, for example, that Resource D can only work on what Resource C delivered to C4 – C Complete – in the previous round."

She added items to the list.

4 Resources (A, B, C, D) performing work in 4 Phases;  balanced capacity
4 Tasks in each Phase – A1, A2, A3, A4 (complete), etc.
4 Resources X 4 Tasks = 16 Tasks for each Project/Job
Resources work only on Projects/Jobs completed by the previous Phase
Reverse Sequence (D-C-B-A)
Game = 24 rounds of D-C-B-A
2 rounds = month;  6 rounds = quarter;  24 rounds = year
1 = effort to complete each Task
4 = effort to complete each Phase (4x1)
4 = avg. work completed/round (1-2-3-4-5-6)

"Each game has 24 rounds of D-C-B-A work.  In terms of operating periods, each round is said to represent a two-week operating period.  I know – there are 26 weekly periods in a year;  24 is close enough, for our purposes.  So, two rounds is said to represent a month of operations, six rounds is said to correlate to a calendar quarter of operations, and all 24 rounds will equate to a year of operations;  easy math.

"We know that every system suffers from variation and uncertainty, the effect of which is to make some tasks take longer than planned.[11]  We know – because of Student Syndrome, Parkinson's Law, and multi-tasking – that tasks will rarely finish early.  In this game, variation is introduced by our six-sided dice;  tasks always take the same amount of effort – in other words, there is no variation in the amount of work, only in the length of time required to complete it.

"Variation is reflected in the probability that a Resource will roll various numbers, reflecting differing amounts of time, every time it performs work.  We find comfort in the false security of achieving an 'average duration', and this game reflects that tendency.  In order to simulate

---

[11] In its new real world, RB Builders will attack the practice of creating job schedules with padded task durations intended to provide a supposed measure of safety;  we discussed that practice in previous sessions.  This game, however, is about managing a production system, not about managing the individual projects/jobs that are in the production system.

that effect, we are going to stipulate one change, which is to eliminate the possibility of any Resource ever rolling a 1;  that leaves the possibility of only rolling a 2-3-4-5-6.

"That will make the average roll a 4.

"That's convenient, because every task takes an effort of 1 to complete it, and there are four tasks in each phase.  Therefore, the total effort required to push a project/job through a phase equals four;  the average amount of work a Resource can complete equals four.  We have enough resource capacity to complete the effort required to get the tasks completed, and to get the jobs/projects moved through the system at the desired pace.

"Let me make sure we are clear:  On average, each round, how much work can each Resource accomplish in their particular phase?"

"Four", replied a superintendent.

"That's right", she said.  "How many tasks does each job/project phase require?"

"Four", said the superintendent.

"Very good", she said.  "We have designed this system with balanced capacity – four phases with four tasks each, four Resources with an average capacity of four;  four phases . . . four tasks . . . four Resources . . . capacity of four . . . balanced capacity.

"A few more rules.  In this game, all the Resources have the ability to work on any project made available to them in their phase.  For example, Resource B can work on any jobs/projects sitting in B1, B2, or B3, but not B4, because any job/project in B4 is complete for that phase;  in addition, Resource B can pull any jobs/projects sitting in A4 – in other words, Resource B can work on any job/project that Resource A completed in the previous round.

"This is what we call the 'standard' game, but be prepared;  there are a number of versions of the standard game, in which we can make changes.  Now – let's talk about how we turn this into a business game."

The intrepid, results-based consultant pulled up a second PowerPoint slide.

## Understanding the Connection

❑ Profitability, cash generation, and economic return are financial outcomes driven by operating performance, productivity, and cycle time.

❑ Operating performance, productivity, and cycle time are linked by the common elements of Throughput, Inventory, and Operating Expense.

"In the second session we did, we talked about the connection between business outcomes and the operating performance that drives it", she said. "We said that profitability, cash generation, and economic return are linked to productivity, inventory turn, and cycle time by the three things – the only three things, operationally speaking – that happen to money."

She pulled up a third PowerPoint slide.

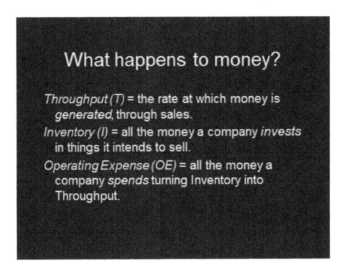

"We said we have these three monetary occurrences. They are the only things that can 'happen to money' from an operational point-of-view. *Throughput* is all of the money RB Builders *generates* through closings. *Inventory* is all of the money the company *invests* in what it intends to transform into that Throughput. *Operating Expense* is everything it *spends* doing it."

"If you recall, we also said financial operating measures have correlating physical operating measures", she said. "Throughput is all the money we generate through closings, but it is also the rate of closings in units; Inventory is all of the money we invest in WIP, but it is also the number of houses under construction; Operating Expense is expressed as the equipment and resources tied to its non-variable costs. As terms of operating performance, Throughput, Inventory, and Operating Expense – closings, work-in-process, and resources – move in response to every operating decision RB Builders makes.

"Unlike financial performance, which is an outcome, operating decisions are happening right now, every day, and we have control over them. We said we have to figure out how to connect the work we do to the outcomes we want, in other words, link operating performance to financial performance, through the terms operating and financial performance hold in common: Throughput, Inventory, and Operating Expense."

She pulled up a fourth PowerPoint slide.

## The Impact on Profitability and Economic Return

Productivity = $Throughput \div Expense$

Inventory Turn = $Throughput \div Inventory$

Net Income = $Throughput - Expense$

ROA = $(Throughput - Expense) \div Inventory$

ROA = $(Net\ Income \div Throughput) \times (Throughput \div Inventory)$

Cycle Time = $(Inventory \div Throughput) \times days$

"There you have it, the key business outcomes and key operating performance drivers, linked by Throughput, Inventory, and Operating Expense.

"Each time we play a game, we will start with a sufficient number of jobs/projects-in-process, spread across the four phases, so that each Resource has enough work to do, and no one is sitting there, twiddling their thumbs", she said.

"Question: The way the standard game is designed, if each Resource just does its job, how many jobs/projects will the system complete each round?"

"Four resources. Four tasks", replied the CFO. "The capacity of each Resource is four; sometimes more, sometimes less, but, on average, capacity of four. WIP that is adequate. So – in a standard game – we should expect to complete one job/project/house every round."

"I'll agree with that number", she replied. "Which means, how many jobs/projects should be completed for the whole game?"

"24 rounds. 24 completed houses", he said.

"Sounds good to me. Throughput for the entire game should be 24 completed jobs/projects", she said. "We have already said that 24 rounds is the equivalent of a year of operation, six rounds equate to a calendar quarter, and two rounds cover a month, which means we should also be generating Throughput – in the form of completed jobs/projects – at the rate of six per quarter and two per month.

"That's the Throughput. What about Inventory?"

"I presume that this game is supposed to simulate an on-going operation, so each phase needs to begin the game with something; I would say each phase needs at least one job/project-in-process", replied a sales representative. "That would be four jobs/projects in-process at the start of the game; otherwise, the Resources won't have enough to do. I think we need to maintain at least that level in WIP throughout the game. Maybe more, I don't know."

"Say we start the game with one job/project in the A4, B4, C4, and D4 tasks", said the intrepid, results-based consultant. "That would be four jobs/projects-in-process as a beginning WIP. If everyone just does an average amount of work – average rolling 4 during the game – four jobs/projects-in-process might be all we need, but let's say that the four Resources, as a team, get to determine how much WIP they carry in the system. There is an unlimited number of ready-to-start jobs/projects in the Start circle, so Resource A can start as many as it wants, and each Resource has been told by management to produce as much work as it individually can; as a matter of fact, that's the way their bonus is structured.

"So – Throughput is expected – budgeted – to be 24, Inventory begins the game at four jobs/projects-in-process and the team figures out how much Inventory they need in order to produce as much as they can. What about Operating Expense? What number would reflect the cost of the capacity required to produce 24 completed jobs/projects per year, per 24 rounds?"

"That's a tough question", replied the VP of Construction.

"What's the value of the Resources?", asked the CEO. "They're what determine your capacity. I'll give you the answer. Resources are worth what they are capable of producing. That's four completed tasks per round. Four resources. Four tasks. It's right there on the flipchart."

*4 Resources X 4 Tasks = 16 Tasks for each Project/Job*

"That's what the system's capacity costs to operate. The system consumes the value of 16 jobs/projects per year, as the cost it incurs to have the capacity to complete 24 jobs/projects.

"The way I look at it, in a standard game, Operating Expense equals 16."

"I'll buy that number", said the intrepid, results-based consultant. "Throughput is budgeted/planned at 24 completed jobs/projects per game, Inventory is planned-but-not-limited to four jobs/projects-in-process, and Operating Expense consumes the value of 16 completed jobs/projects.

"I have a few more questions", she said, pointing at the slide. "How would you calculate the business outcomes and operating performance drivers? Let's start with operating performance.

"How would you calculate Productivity?"

"Productivity = Throughput ÷ Expense, so I suppose it would be whatever 24 ÷ 16 is", replied a superintendent. "1.5. It's kind of an abstract number, but it's a useful reference."

The intrepid, results-based consultant added the calculation to the flipchart.

*Productivity = Throughput ÷ Expense*
*Productivity = 24 ÷ 16*

Productivity = 1.5

"Okay.  What about Inventory Turn?", she asked.

"Inventory Turn = Throughput ÷ Inventory", said CFO.  "24 ÷ 4 = 6.  In a standard game, we would say we should have a 6x turn;  we should turn our inventory six times a year, six times during the course of the game."

The intrepid, results-based consultant added to the flipchart.

Inventory Turn = Throughput ÷ Inventory
Inventory Turn = 24 ÷ 4
Inventory Turn = 6

"That's right.  RB Builders strives to keep land and lot inventory off its Balance Sheet, so its Inventory Turn is essentially its Asset Turn, which makes it one of the two components used to calculate Return on Assets", she said.  "Inventory Turn is also an indicator of velocity, and the reciprocal of cycle time.

"Speaking of which, what about cycle time – how do we calculate the duration of the four phases in the process?"

"There's a difference between how you *calculate* duration and how you *measure* duration", said the senior superintendent.  "The equation used to calculate cycle time is (Inventory ÷ Throughput) x $Days_n$.  In a typical calendar-based cycle time calculation, the number of days – the $Days_n$ part – is the number of days in the period for which cycle time is being calculated.

"In this game, it's a bit different, because we are dealing with the number of rounds as the duration, not the number of days;  I would say we need to calculate the number of rounds, relate it to rounds per calendar event, and then the number of days we associate with that event.

"In a standard game, projected Throughput is 24 and projected Inventory is four.  In order to calculate cycle time for the entire game, which would also be the equivalent of the calculation for the entire year, it would be 4 ÷ 24;  take that quotient – 0.167 – and multiply it by 24.  That would be 0.167 x 24 = 4.  The expected cycle time would be four rounds.

Cycle Time = (Inventory ÷ Throughput) x $Days_n$
Cycle Time = (4 ÷ 24) x 24
Cycle Time = 0.167 x 24
Cycle Time = 4 rounds

"Each round represents two weeks, so four rounds represents eight weeks;  we don't typically use a weeks-per-month calculation, but we'll use four weeks as the standard for a month, which makes the calculated cycle time two months.  For ease-of-calculation, we say a year is 360 days, which makes 30 days standard for a month;  a two-month duration is 60 days."

The senior superintendent reached across the conference table and borrowed the intrepid, results-based consultant's HP-12C.

"Coelacanth agrees", he said, after punching in the numbers. "60 days is correct, and it corresponds to the reciprocal of the 6x Inventory Turn we had calculated. I have to say, 60 day cycle times and 6x turns don't sound much like RB Builders' current operating performance. We need more Inventory."

"More cowbell", said the superintendent.

"Why?", asked the intrepid, results-based consultant.

"Yeah, why?", echoed the VP of Sales. "If all of you guys with the Silverados and F-150s just do your jobs – roll an average of 4 – you 'have a sufficient number jobs/projects-in-process to meet your targeted Throughput with a finite and controlled amount of capacity'. I think that's how she says it.

"After all, my sales team and I have personally arranged for a limitless number of ready-to-start jobs/projects to always be sitting in the Start circle, so that Resource A can start as many as it wants."

"That's why we pay you, Rainmaker", said the CEO.

"Back to the question", said the intrepid, results-based consultant. "You say you need more Inventory. Why?"

"Very simple", said the senior superintendent. "Variation. We have a sufficient number of jobs/projects-in-process, if everything goes right. Except, it won't go right. Murphy will make its customary appearance, and we have no protection from it. The Law of Variability Buffering, correct?

"And – yes – I do know we need to do whatever we can to reduce variation and uncertainty", he said, "but if we don't provide a buffer, the system will take matters into its own hands.

"We know it will be a cold day in Hell when we get additional capacity, in the form of more Resources; that would push Operating Expense above 16, and that's just not going to happen. So, if we don't get additional jobs/projects-in-process – more cowbell – the system will simply add duration, longer cycle time; and, longer cycle time will reduce the number of completed jobs/projects.

"That will reduce Throughput, both physical and financial."

"You make a valid point, actually, a couple of them", said the intrepid, results-based consultant. "Regarding your immediate point, concerning the sufficiency of Inventory. I'm sure you recall the discussion of Necessary WIP in an earlier session. You are correct about the ways in which a system will protect itself from variation. It's a valid concern. However, we've already stipulated that – at least in the initial game – the Resources determine how much WIP they carry in the system. And, we'll see how that plays-out.

"Regarding the difference between calculated cycle time and measured cycle time, there are – as you indicate – two ways to determine it.  On one hand, *measured cycle time* is the average, or mean duration, of a set of jobs;  its value lies in examining the forensics of individual jobs, in order to eliminate the causes of problems, variation, and waste.  It's where we use kaizen, PDCA.  The value of measuring average cycle time lies in the selection of individual jobs to examine, because they are outside the lines of statistical control.

"On the other hand, *calculated cycle time* is the relationship between completed jobs and work-in-process, two values that relate to Throughput and Inventory, respectively;  the value of calculated cycle time is in providing a picture of the condition and performance of the entire system.

"In terms of what is actionable, *measured* cycle time is actionable at the level of every future job based on the forensics of past jobs, while *calculated* cycle time is actionable in ways that affect the performance of the entire system.

"That covers the operating performance measures", she said.  "What about the measures of business, or financial, outcomes that are driven by that operating performance?  If you were playing a standard game, how would you measure Net Income?"

"That's a question I would like to answer", said the CFO, moving to the flipchart.

"By focusing on Throughput, we are purposely taking Revenue and Cost of Sales out of the equation.  I'm okay with that, because this is a production game;  the game highlights velocity more than it does margin.

"When you think about it, each completed job/project generates Revenue that is depleted by the direct, variable costs incurred to produce it.  Job Income and Job Costs aren't moved to the Income Statement until the job/project is completed and closed-out.  Until then, those costs live on the Balance Sheet, which is why we consider Throughput synonymous with Gross Income, or, more correctly, Contribution Margin."

The CFO wrote as he spoke.

$$\text{Net Income} = \text{Throughput - Operating Expense}$$
$$\text{Net Income} = 24 - 16$$
$$\text{Net Income} = 8$$
$$\text{Net Margin} = \text{Net Income} \div \text{Throughput}$$
$$\text{Net Margin} = 8 \div 24$$
$$\text{Net Margin} = 33\%$$

"Net Income is *Throughput - Operating Expense*.  We have already established that we expect to complete 24 jobs/projects over the course of a game, and that's our Gross Income, with Cost of Sales already converted, or transformed.  In the process, the system will consume the value of 16 jobs/projects, as the cost it incurs for the resource capacity to complete those 24 jobs/projects.

"24 - 16 = 8.

"So – in a standard game, we expect to have a Net Income of eight jobs/projects, and the Net Income Margin would be calculated as Net Income divided by Throughput, which would be:

"8 ÷ 24 = 33%."

"Thank you", said the intrepid, results-based consultant.  "That was good.  This calculation is also Return on Sales, whenever we use the DuPont formula for calculating Return on Assets, or ROA.  Now – do you recall the component of ROA that Return on Sales represents?"

"It represents the margin component", answered a sales representative.

"That's right.  And, what represents the other component of ROA?", asked the intrepid, results-based consultant.  "What represents velocity?"

"Asset Turn", said a superintendent.  "Although, since we are using Inventory Turn, let's call ROA what it is:  Return on Invested Assets."

"Yes – we can call it ROIA", said the intrepid, results-based consultant, writing on the erasable board.  "Learn this progression."

$$ROIA = (Net\ Income \div Throughput) \times (Throughput \div Inventory)$$
$$ROIA = Return\ on\ Sales \times Inventory\ Turn$$
$$ROIA = Margin \times Velocity$$

"Game 1;  first game, standard game", she said.  "What is our budgeted ROIA?"

One of the sales representatives walked to the erasable board and wrote:

$$ROIA = (8 \div 24) \times (24 \div 4)$$
$$ROIA = .33 \times 6$$
$$ROIA = 1.99$$
$$ROIA = 199\%$$

"In Game 1, we expect an economic return, a Return on Invested Assets, of almost 200%. Those numbers don't exactly look like a homebuilding operation, but they make perfect sense, based on the game rules.  Four jobs/projects 'invested', and we expect to complete 24 jobs/projects;  ROIA of 200%."

"Okay, let's play it", said the intrepid, results-based consultant.  "Game 1 is a production system with balanced capacity.  Focus on learning and on understanding the principles."

GAME 1

In Game 1, the assignments of Resource roles all went to superintendents. The CFO kept score during the game, and tallied the scorecard . At the end of 24 rounds, the intrepid, results-based consultant asked, "How did we do?"

"Not so good", replied the CFO.

"Give me the results", she said, moving to the erasable board and clearing space for what would become a large tracking table.

"Well, for starters, we only completed 19 jobs/projects, not the 24 jobs/projects that we budgeted", the CFO said. "So, our Net Income was three jobs/projects, not the eight jobs/projects budgeted, which is a Net Income Margin of 15.8%, not the 33% we were talking about, and, that was on the reduced actual; if we calculated it on budget, Net Margin would have only been 12.5%.

"ROIA was a measly 38%, not the anticipated 200%."

"Keep going", she said.

"The disappointing ROIA wasn't just a function of reduced Net Income Margin", he said. "We were supposed to turn our Inventory six times; instead of a 6x Inventory Turn, we had a 2.4x."

"Why did that happen?", asked the intrepid, results-based consultant.

"In part, because we closed fewer jobs/projects than expected", he replied. "The other factor was the amount of Inventory we carried. We started the game with four jobs/projects-in-process, but it kept increasing as it went on; the final-third of the game, we were carrying 10 or 11 jobs/projects-in-process. For the game, we averaged eight jobs/projects-in-process."

| | G-1 | G-2 | G-3 | G-4 | G-5 | G-6 | G-7 | G-8 |
|---|---|---|---|---|---|---|---|---|
| Description<br>Std. = 4<br>Smart = 4<br>Strong = 5 | Balanced, 4 Std., unlimited WIP | | | | | | | |
| | | | | | | | | |
| Budgeted | | | | | | | | |
| T | 24 | | | | | | | |
| I | 4 | | | | | | | |
| OE | 16 | | | | | | | |
| P-Rate | 1.5 | | | | | | | |
| Inv. Turn | 6.0x | | | | | | | |
| CT (days) | 60 | | | | | | | |
| NI | 8 | | | | | | | |
| NI % | 33% | | | | | | | |
| ROIA | 200% | | | | | | | |
| | | | | | | | | |
| Actual | | | | | | | | |
| T | 19 | | | | | | | |
| I | 8.0 | | | | | | | |
| OE | 16 | | | | | | | |
| P-Rate | 1.2 | | | | | | | |
| Inv. Turn | 2.4x | | | | | | | |
| CT (days) | 150 | | | | | | | |
| NI | 3 | | | | | | | |
| NI % | 16% | | | | | | | |
| ROIA | 38% | | | | | | | |
| | | | | | | | | |
| Resource A | 4.6 | | | | | | | |
| Resource B | 3.5 | | | | | | | |
| Resource C | 4.0 | | | | | | | |
| Resource D | 3.9 | | | | | | | |
| | | | | | | | | |

"And, what was your calculated cycle time?", she asked.

"We completed 19 jobs/projects, that was our Throughput, our production. We carried an average of eight jobs/projects-in-process", said the CFO, moving to the flipchart. "Based on Little's Law, our calculated duration was 10 rounds, which is the equivalent of five months, or 150 days."

$$CT = (WIP \div T) \times Days$$
$$CT = (8 \div 19) \times 24$$
$$CT = 10 \text{ periods}$$
$$CT = (10 \div 2) \times 30$$

CT = 150 Days

"What was your cycle time supposed to be?", she asked.

"It was supposed to be four rounds, which would have been equivalent to two months, or 60 days", said the CFO. "And, before you ask, we thought our productivity rate – calculated as Throughput ÷ Operating Expense – would be 1.5, but the actual productivity rate was only 1.2.

"Since the cost of a Resource is a function of its expected capacity to produce, Operating Expense came in exactly where it was expected; it was the less-than-expected Throughput that was the problem. You'll notice that I rounded some of the numbers; I also included the average work available from each Resource, based on the average dice roll it produced over the 24 periods."

"Anything else?", asked the intrepid, results-based consultant, pointing to the table. "What else stands-out to you in Game 1?"

"For one thing, twice as much inventory as we agreed would be necessary", said one of the superintendents. "And, a ton of variation between the amounts of work delivered by the four Resources."

"Meaning what?", asked the CEO.

"In Game 1, each Resource was left to determine how many jobs/projects it would have in its phase at any point in time, somewhat determined by how much work it could perform", replied the superintendent. "The point is, there was nothing that encouraged any of the Resources to coordinate their Inventory level decisions, or even which jobs/projects they would expend their effort; in fact, the Resources were told to produce as much as they could, because that is what their bonuses were calculated on.

"If you calculate the amount of variation from what the average roll should have been – which was a 4 – you have Resource A who was 15% stronger and you have Resource B who was almost 13% weaker; that's close to 30% variation, and it occurred between two side-by-side Resources that we had counted on to produce the same amount of work.

"Also – Resource A produced more than what was needed, Resources B and D produced less than what was needed, and Resource C produced the amount of work we all agreed we needed."

"What else?", asked the intrepid, results-based consultant.

"Throughput", said one of the sales representatives. "The periodic variation in the number of completed jobs/projects is significant. We had reasoned that one completed job/project per period – one job/project per period moved into D4 – is what we needed, in order to achieve budgeted Revenue. There were three periods in which we moved two jobs/projects into D4, but there were also eight periods in which we failed to complete even one job/project."

"Alright, good stuff", she said. "Play the next game."

133

GAME 2

Game 2 was another standard game.  The only change installed for Game 2 was to limit system-wide Inventory to a total of four jobs/projects-in-process – a Max WIP rule of four;  by comparison, Game 1 had no limits on Inventory.

"Tell us about the results for Game 2", said the intrepid, results-based consultant, moving back to the tracking table on the erasable board.

"Game 2 was better in some ways, worse in others;  overall, it was worse", said the CFO. "We only completed 16 jobs/projects, not the 24 jobs/projects that we budgeted, not even the 19 that we completed in Game 1."

"We achieved breakeven, if you call that an achievement.  There was zero Net Income, certainly no profit.  Our completed jobs/projects absorbed our overhead – i.e., our Operating Expense – but nothing above, definitely not the eight jobs/projects budgeted.  That resulted in a Net Income Margin of 0.0%, also not the 33% we budgeted;  and, zero Net Income means zero economic return, so – instead of earning an ROIA of 200% – it was zero, too."

"Big Fat Zero.  What about your operating measures?", she asked.

"Our non-existent ROIA was a function of our equally non-existent Net Income Margin", he said.  "We consistently maintained WIP at four jobs/projects-in-process, so our Inventory Turn was 4x, which was better than the 2.4x we had in Game 1.  It would have been the 6x that we projected, if our Throughput hadn't been eight jobs/projects short of expectations."

| | G-1 | G-2 | G-3 | G-4 | G-5 | G-6 | G-7 | G-8 |
|---|---|---|---|---|---|---|---|---|
| **Description**<br>Std. = 4<br>Smart = 4<br>Strong = 5 | Balanced, 4 Std., unlimited WIP | Balanced, 4 Std., Max WIP = 4 | | | | | | |
| | | | | | | | | |
| **Budgeted** | | | | | | | | |
| T | 24 | 24 | | | | | | |
| I | 4 | 4 | | | | | | |
| OE | 16 | 16 | | | | | | |
| P-Rate | 1.5 | 1.5 | | | | | | |
| Inv. Turn | 6.0x | 6.0x | | | | | | |
| CT (days) | 60 | 60 | | | | | | |
| NI | 8 | 8 | | | | | | |
| NI % | 33% | 33% | | | | | | |
| ROIA | 200% | 200% | | | | | | |
| | | | | | | | | |
| **Actual** | | | | | | | | |
| T | 19 | 16 | | | | | | |
| I | 8.0 | 4.0 | | | | | | |
| OE | 16 | 16 | | | | | | |
| P-Rate | 1.2 | 0 | | | | | | |
| Inv. Turn | 2.4x | 4.0x | | | | | | |
| CT (days) | 150 | 90 | | | | | | |
| NI | 3 | 0 | | | | | | |
| NI % | 16% | 0% | | | | | | |
| ROIA | 38% | 0% | | | | | | |
| | | | | | | | | |
| Resource A | 4.6 | 4.1 | | | | | | |
| Resource B | 3.5 | 4.5 | | | | | | |
| Resource C | 4.0 | 4.1 | | | | | | |
| Resource D | 3.9 | 3.8 | | | | | | |
| | | | | | | | | |

"Really?", asked the intrepid, results-based consultant.

"I don't think so", said the VP of Construction.  "I think we kept Inventory too low, and we starved the system.  That level of Throughput, projected with that level of Inventory, was an illusion.  In theory, four jobs/projects-in-process should be sufficient, but it doesn't provide any buffer against variation.  A system will protect itself from variation with some combination of additional work-in-process, longer duration, and/or excess capacity. In this game, there was certainly no buffer of additional jobs/projects-in-process."

"Interesting observation.  What about the other buffers?  What was cycle time?"

"90 days", answered one of the superintendents, walking to the flipchart.

$$CT = (WIP \div T) \times Days$$
$$CT = (4 \div 16) \times 24$$
$$CT = 6 \; periods$$
$$CT = (6 \div 2) \times 30$$
$$CT = 90 \; Days$$

"We consistently maintained a level of four jobs/projects-in-process, and we completed 16 jobs/projects", the superintendent explained. "That's a calculated duration of six periods/rounds, or 90 days. That is too small a difference from the projected 60 day cycle time to provide much of a buffer, so I think the system is mostly protecting itself with excess capacity, by failing to produce the desired amount of Throughput with the capacity it has available, all in response to variation in the amount of work the Resources deliver."

"Yes – adequate capacity becomes excess capacity, if the system won't utilize it", said the intrepid, results-based consultant. "How much variation?"

"If you calculate it from the average roll of 4 that we should have had, Resource B was almost 13% stronger and Resource D turned out to be 5% weaker;  that's an 18% variation.

"A, B. and C all got the job done, averaging rolls better than 4.0, but D didn't;  D only averaged 3.8 rolls per period. Moreover, D was batting cleanup, which is the worst position for that kind of performance.

"That doesn't sound like much, but 50% of D's rolls were either a 2 or a 3, which means that D had no chance of moving a job/project in C4 (complete) to D4 (complete) on any of those rolls;  with the reduced availability of jobs/projects-in-process, D was bound to strand jobs/projects on-base in its phase. Game 2 was set-up so that rolling 2s and 3s would kill it, and rolling 5s and 6s would be wasted;  there was no Inventory buffer.

"There was one period/round in which D completed two jobs/projects, but there were nine periods/rounds in which it didn't complete any jobs/projects. With this much variation in the amount of work Resources performed, and such a lean level of inventory, we didn't have a prayer."

"Okay, then let's begin to deal with that variation", said the intrepid, results-based consultant. "In our game, how would we reflect a successful effort to reduce variation?  How would less variation manifest itself?"

"Variation is reflected in the dice rolls", said one of the superintendents who had been one of the Resources in the first two games. "Any reduction in variation would, therefore, have to be reflected in the rolls."

"What if we changed the probability?", asked the CFO. "What if we kept the same average roll – 4 – but we reduced the variation?"

"The so-called 'standard' Resource is a dice that has a 2-3-4-5-6 roll possibility, because we had eliminated the possibility of a 1", said the intrepid, results-based consultant. "That gives the standard dice an average roll of 4, and gives the standard Resource an average workrate/capacity of 4. Let me introduce to you our new, smarter Resources."

Standard:  4 = avg. work completed/round (1-2-3-4-5-6)
Smart:  4 = avg. work completed (1 or 2 = 3; 3 or 4 = 4; 5 or 6 = 5)

"These 'smart' Resources can roll anything on a dice face, 1-2-3-4-5-6; a roll of 1 or 2 = 3, a roll of 3 or 4 = 4, and a roll of 5 or 6 = 5. As you can see, the standard Resource and the 'smart' Resource both have average roll/workrate/capacities of 4, but the standard Resource has a probability distribution of 20%-20%-20%-20%-20%, while the 'smart' Resource has a distribution of 33%-33%-33%.

"That means, while the standard Resource had a 60% probability for a workrate of 4 or higher, the 'smart' Resource has a 67% probability; it can't produce as much at the top-end, by rolling a 6, but it isn't in danger of producing at the bottom-end, by rolling a 2, either.

"3-4-5, instead of 2-3-4-5-6.

"Remember how the cost of a production system's capacity is calculated in these games. Since it has the same workrate/capacity, a 'smart' Resource has exactly the same cost as a standard Resource; so – if the production system has four 'smart' Resources, it has exactly the same Operating Expense as a production system with four standard Resources."

Operating Expense = #Resources x workrate-capacity
Operating Expense = 4 x 4
Operating Expense = 16

"Same workrate/capacity, same cost, more stability, less variation."

GAME 3

Game 3 was a standard game, played with 'smart' Resources. Once again, it was a system with balanced capacity across all four Resources, who each performed the same amount of work – the same number of tasks – in each phase, over 24 periods/rounds.

The other change in Game 3 was a mandated upper and lower range of jobs/projects-in-process – in other words, a minimum and maximum range of Inventory/WIP; six jobs/projects-in-process minimum, eight jobs/projects maximum.

"How did we do?", asked the intrepid, results-based consultant.

"Better", said the CFO.

"We completed 22 jobs/projects, and we kept our jobs/projects-in-process within the specified range. At 22 jobs/projects completed, Throughput wasn't quite as high as anticipated, so our Net Income wasn't either; it came in at six jobs/projects, instead of the budgeted eight jobs/projects.

"That produced a 27% Net Income Margin. Inventory Turn was 3.1x, just under the anticipated 3.4x, which itself was a lower anticipated turn, due to the larger number of jobs/projects-in-process that we decided to carry, in order to protect/buffer the system from variation."

"Compared to previous games, there was considerably less variation in the effective Resource workrate", said the CFO. "The variation between the strongest and weakest Resource produced a variance of about 8% in Game 3, compared to variances of 18% and 30% in Games 1 and 2."

"What effect did the move to a mandated minimum and maximum range of Inventory/WIP have?", asked the CEO.

"I think it was a good move", replied the VP of Construction. "The imposition of a range of jobs/projects-in-process provided some needed discretion and flexibility, while the level of six-to-eight seemed right, given the capacity of the system and the reduced amount of variation; without the reductions in variation, we would have needed higher levels of Inventory/WIP.
"There was a gradual, but fairly steady increase from the lower WIP range to the higher WIP range as the game progressed. We were consistently at Max WIP during the final-third of the game."

| | G-1 | G-2 | G-3 | G-4 | G-5 | G-6 | G-7 | G-8 |
|---|---|---|---|---|---|---|---|---|
| Description<br>Std. = 4<br>Smart = 4<br>Strong = 5 | Balanced, 4 Std., unlimited WIP | Balanced, 4 Std., Max WIP = 4 | Balanced, 4 Smart, WIP = 4 to 6 | | | | | |
| | | | | | | | | |
| **Budgeted** | | | | | | | | |
| T | 24 | 24 | 24 | | | | | |
| I | 4 | 4 | 6-8 | | | | | |
| OE | 16 | 16 | 16 | | | | | |
| P-Rate | 1.5 | 1.5 | 1.5 | | | | | |
| Inv. Turn | 6.0x | 6.0x | 3.4x | | | | | |
| CT (days) | 60 | 60 | 90-120 | | | | | |
| NI | 8 | 8 | 8 | | | | | |
| NI % | 33% | 33% | 33% | | | | | |
| ROIA | 200% | 200% | 112% | | | | | |
| | | | | | | | | |
| **Actual** | | | | | | | | |
| T | 19 | 16 | 22 | | | | | |
| I | 8.0 | 4.0 | 7.0 | | | | | |
| OE | 16 | 16 | 16 | | | | | |
| P-Rate | 1.2 | 0 | 1.4 | | | | | |
| Inv. Turn | 2.4x | 4.0x | 3.1x | | | | | |
| CT (days) | 150 | 90 | 115 | | | | | |
| NI | 3 | 0 | 6 | | | | | |
| NI % | 16% | 0% | 27% | | | | | |
| ROIA | 38% | 0% | 84% | | | | | |
| | | | | | | | | |
| Resource A | 4.6 | 4.1 | 3.9 | | | | | |
| Resource B | 3.5 | 4.5 | 3.9 | | | | | |
| Resource C | 4.0 | 4.1 | 4.2 | | | | | |
| Resource D | 3.9 | 3.8 | 4.0 | | | | | |
| | | | | | | | | |

"True – there was less variation between the output of the four Resources, but there was still the effect of variation in the Throughput of the system", said another superintendent who had been a Resource. "There were four rounds/periods in which we completed two jobs/projects, but there six other rounds/periods in which failed to complete a job/project."

"Okay", said the intrepid, results-based consultant. "Game 3 was the best result, so far. However, none of the three scenarios we have employed in the first three games has produced the results we expected. We've changed the level of Inventory, we have reduced Resource variation.

"It must be so terribly frustrating, dealing with such a sense of ineptness", she said, smiling. "Any ideas for what we should do about it?"

"More cowbell", said the superintendent.

"The universal solution anytime you boys need more of anything", she said. "First, it was resolving resource conflict. Then, it was about Inventory. What is it this time? More capacity?"

"We balanced the capacity", said the superintendent. "We've adjusted Inventory. We've reduced variation; we've used 'smart' Resources. Sometimes, you just need more capacity to get more work done, to complete more jobs/projects. Sometimes, you need more, to do more.

"That's what we've always done; more-for-more.

"More cowbell."

"Alright then, more resource capacity it is", conceded the intrepid, results-based consultant. "We could resort to adding more bodies – to adding more of the same strength Resources on each phase – or, we can replace all of the existing Resources with better, stronger, more powerful Resources.

"What would better, stronger, more powerful Resources look like?"

"They would need to be capable of rolling more than the existing average roll", said one of the other superintendents. "How about an average roll of 5? That way, all of the Resources should be able to roll higher than what they need. Each phase has four tasks, so anything above a roll of 4 should work. What about a Resource that can average a roll of 5, with the same, improved variance control we had with the smart Resources?"

"As we already know, the standard Resource has a 2-3-4-5-6 dice roll possibility, because we had eliminated the possibility of rolling a 1", the intrepid, results-based consultant answered. "The 'smart' Resource can roll anything on the dice face, 1-2-3-4-5-6, with a roll of 1 or 2 = 3, a roll of 3 or 4 = 4, and a roll of 5 or 6 = 5. The standard Resource and the 'smart' Resource both have average workrate/capacities of 4.

"Based on your suggestion, there's a new strain of 'strong' Resources available", she said. "Better; stronger; more powerful.

"You can think of them as 'smart' Resources on steroids."

Standard: 4 = avg. work completed/round (1-2-3-4-5-6)
Smart: 4 = avg. work completed (1 or 2 = 3; 3 or 4 = 4; 5 or 6 = 5)
Strong: 5 = avg. work completed (3 or 4 = 4; 2 or 5 = 5; 1 or 6 = 6)

"As with 'smart' Resources, we will let 'strong' Resources roll anything on the dice face, 1-2-3-4-5-6", she said. "However, with 'strong' Resources, a roll of 3 or 4 = 4; a roll of 4 or 5 = 5; a roll of 1 or 6 = 6. That gives 'strong' Resources an average roll/workrate/capacity of 5, while the standard Resource and the 'smart' Resource have an average roll/workrate/capacity of 4.

"We have given 'strong' Resources the same 33%-33%-33% probability distribution as 'smart' Resources, an improvement over the 20%-20%-20%-20%-20% of a standard Resource's distribution. So – while the standard Resource has a 60% probability of 4 or higher, and the 'smart' Resource has a 67% probability of 4 or higher – the 'strong' Resource has a 67% probability for a workrate of 5 or higher.

"4-5-6, instead of either 3-4-5, or 2-3-4-5-6.

"Because of how the cost of the system's capacity is calculated, standard Resources and 'smart' Resources have the same cost, because they have the same workrate/capacity", she said. "And, because a 'strong' Resource has a higher workrate/capacity, logic and market sense says it is going to incur a proportionally higher cost.

"Whereas, either a standard Resource or a 'smart' Resource incurs a cost of four jobs/projects, a 'strong' Resource incurs a cost of five jobs/projects; so, if the system has four 'strong' Resources, it will incur a higher total cost – higher Operating Expense – of 20 jobs/projects per game, compared with a cost of 16 jobs/projects incurred for a system with all standard or 'smart' Resources.

"Like this", she said.

Operating Expense = #Resources x workrate-capacity
Operating Expense = 4 x 5
Operating Expense = 20

"Higher workrate/capacity, higher cost, same stability."

The CEO looked at the four superintendents who were the Resources in the first three games, smiled, and said, "It looks like you just got fired."

"As I've been diligent to remind everyone, every time I've had an opportunity", said the VP of Sales, "you are fortunate that my sales team and I have personally arranged for that limitless number of ready-to-start jobs/projects, sitting in the Start circle, so Resource A can start as many as it wants. Because – we wouldn't want to miss out on all of that additional Revenue, would we?"

"You are so right, Rainmaker", said the intrepid, results-based consultant. "In the previous games, we had the capacity to complete one job/project per period, which should have produced 24 completed jobs/projects per game – not that we ever got there. Now, we should be able to complete more.

"Question is: How many more?"

"I think we should be able to complete six additional jobs/projects per game", said one of the sales representatives. "Four phases; four Resources; four tasks per phase; only, now, the capacity of each Resource is five; that's 25% more capacity. If we were supposed to be able to complete 24 jobs/projects before, now we should be able to complete 30 jobs/projects.

"Here," said the sales representative, going to the erasable board. "Do the math."

24 completed jobs/projects x 1.25 = 30 jobs/projects

"So, how will the increased capacity – with its attendant higher cost and promise of increased Throughput – affect Net Income and economic return?", asked the intrepid, results-based consultant.

"We expect to complete 30 jobs/projects", said the sales representative. "That's now our Gross Income, with Cost of Sales already converted, or transformed. In the process, the system will now consume the value of 20 jobs/projects per year, as the cost – the Operating Expense – it incurs for having the capacity to complete 30 jobs/projects."

Net Income = Throughput – Operating Expense
Net Income = 30 – 20
Net Income = 10

"In an otherwise-standard game played with 'strong' Resources, we can expect a Net Income of 10 jobs/projects; Net Income Margin would be 33%."

Net Income Margin = Net Income ÷ Throughput
Net Income Margin = 10 ÷ 30
Net Income Margin = 33%

"Very good, rep-san", said the intrepid, results-based consultant. "As we have previously noted, the calculation for Net Income also serves as the Return on Sales component, when we do a DuPont formula calculation for Return on Assets. So – calculate the anticipated Inventory Turn, and then give us the projected Return on Invested Assets."

"I don't know about the rest of you, but I recommend we restrict Inventory to eight jobs/projects-in-process, minimum and maximum", said the sales representative. "If we do that, Inventory Turn should be 3.8x. We've already said expected Return on Sales is 33%.

"Here, like this", she said, pointing, then changing and adding, to a previous set of calculations on the erasable board. "In Game 3, we expect an economic return – a Return on Invested Assets – of 125%."

ROIA = (Net Income ÷ Throughput) x (Throughput ÷ Inventory)
ROIA = Return on Sales x Inventory Turn

$ROIA = Margin \times Velocity$

$ROIA = (10 \div 30) \times (30 \div 8)$

$ROIA = .33 \times 3.8$

$ROIA = 1.25$

$ROIA = 125\%$

"We have a recommendation that Inventory be limited to eight jobs/projects-in-process, minimum/maximum. As a team, how do you think the decision to install 'strong' Resources with a higher workrate or higher capacity will affect WIP?", asked the intrepid, results-based consultant. "In other words, how many jobs/projects-in-process do you think the system is going to require, in order to fully-utilize its additional capacity?"

"Good question", replied the VP of Construction. "One of the lessons we have learned from this entire series on production principles has involved the interdependent nature of the parts of a system.

"If each Resource is going to have more capacity than the amount of work there is to do, then it stands to reason that the minimum Inventory requirement we initially calculated – one job/project-in-process in each of the four phases – is no longer adequate. It's a production system with balanced capacity. If we don't increase work-in-process throughout the system, we risk starving it, wasting the capacity, and reducing Throughput.

"At the risk of occasionally having too much Inventory, a minimum of eight jobs/projects-in-process seems about right, two in each phase."

"It doesn't appear that we have a choice", said the CEO, ominously. "The system just got bigger".

GAME 4

Game 4 was a standard game, played with 'strong' Resources. Once again, it was a system with balanced capacity across all four Resources, who each performed the same amount of work – the same number of tasks – in each phase, over 24 periods/rounds.

The other change in Game 4 was the mandated upper and lower range of jobs/projects-in-process – in other words, a minimum and maximum range of Inventory/WIP; in Game 4, the specified minimum and maximum amount of Inventory was the same – eight jobs/projects-in-process.

"Results?", asked the intrepid, results-based consultant.

"Much better", replied the CFO. "We played this game twice, with very similar results; we scored both games, but we're only reporting the lowest result. You can clearly see the benefit of Resources that are both smart and strong, even though the 'strong' Resources come at a higher cost.

"Throughput was just below expectations, at 29 completed jobs/projects. We knew we were incurring the additional cost of 'strong' Resources, in the form of higher Operating Expense, but our Net Income was still nine completed jobs/projects, giving us a Net Income Margin of 31%, best so-far.

"Our Inventory was rock-steady. We began the game with eight jobs/projects-in-process, and we ended the game with eight jobs/projects-in-process; in fact, our total WIP never varied throughout 24 rounds/periods. That gave us an Inventory Turn of 3.6x, slightly lower than the anticipated 3.8x, due entirely to lower Throughput."

ROIA = .31 X 3.6
ROIA = 1.12
ROIA = 112%

"That gave us an ROIA of 112%".

| | G-1 | G-2 | G-3 | G-4 | G-5 | G-6 | G-7 | G-8 |
|---|---|---|---|---|---|---|---|---|
| Description<br>Std. = 4<br>Smart = 4<br>Strong = 5 | Balanced, 4 Std. unlimited WIP | Balanced, 4 Std. Max WIP = 4 | Balanced, 4 Smart WIP = 4 to 6 | Balanced, 4 Strong WIP = 8 Min/Max | | | | |
| | | | | | | | | |
| **Budgeted** | | | | | | | | |
| T | 24 | 24 | 24 | 30 | | | | |
| I | 4 | 4 | 6-8 | 8 | | | | |
| OE | 16 | 16 | 16 | 20 | | | | |
| P-Rate | 1.5 | 1.5 | 1.5 | 1.5 | | | | |
| Inv. Turn | 6.0x | 6.0x | 3.4x | 3.8x | | | | |
| CT (days) | 60 | 60 | 90-120 | 96 | | | | |
| NI | 8 | 8 | 8 | 10 | | | | |
| NI % | 33% | 33% | 33% | 33% | | | | |
| ROIA | 200% | 200% | 112% | 125% | | | | |
| | | | | | | | | |
| **Actual** | | | | | | | | |
| T | 19 | 16 | 22 | 29 | | | | |
| I | 8.0 | 4.0 | 7.0 | 8.0 | | | | |
| OE | 16 | 16 | 16 | 20 | | | | |
| P-Rate | 1.2 | 0 | 1.4 | 1.5 | | | | |
| Inv. Turn | 2.4x | 4.0x | 3.1x | 3.6x | | | | |
| CT (days) | 150 | 90 | 115 | 99 | | | | |
| NI | 3 | 0 | 6 | 9 | | | | |
| NI % | 16% | 0% | 27% | 31% | | | | |
| ROIA | 38% | 0% | 84% | 112% | | | | |
| | | | | | | | | |
| Resource A | 4.6 | 4.1 | 3.9 | 4.9 | | | | |
| Resource B | 3.5 | 4.5 | 3.9 | 5.1 | | | | |
| Resource C | 4.0 | 4.1 | 4.2 | 5.0 | | | | |
| Resource D | 3.9 | 3.8 | 4.0 | 4.8 | | | | |
| | | | | | | | | |

"We calculated our cycle time at 99 days, in line with the 3.6x Inventory Turn", he said. "Productivity was on-budget, at 1.5.

"In terms of capacity utilization, the effort to reduce process variation by utilizing smart-but-now-strong Resources continued to pay rewards. There was only a 6% variance between most and least productive Resource; all four Resources were within 4% of the average 'strong' roll of 5.

"What continues to plague us a bit is the percentage of diminished work. Overall, 38% of the rolls were below-average, which should have occurred only one-third of the time.

"There were also troubling differences between Resources: Almost 50% of Resource A's and Resource D's rolls were a 4, compared with 25% for Resource B and 33% for Resource C. Basically, we didn't get enough production out of either our lead-off batter or our clean-up hitter.

"But – overall – Game 4 was our best performance.

"I think that's as good as it gets."

"No, it isn't", said the intrepid, results-based consultant.

"You make a worthwhile distinction between the use of 'smart' Resources and the use of 'strong' Resources", she said. "The terms 'smart' and 'strong' are not mutually-exclusive; the Resources we use can be either smarter or stronger, they can be both smarter and stronger, and they can be neither.

"The distinction is that we are able to utilize 'smart' Resources because we have attacked variation in the system, but smarter resources don't add to our Operating Expense, and we don't get any more work out of them. On the other hand, we are able to employ 'strong' Resources, because we either want or need to add capacity. In either event, we only get to use 'strong' Resources because we are willing to pay more to have them.

"In Game 1, we saw the effect variation has on standard Resources in a system that seemed to be adequately designed; we agreed the capability was there, but variation prevented the system from performing the way it could have.

"In Games 2 and 3, we caught a glimpse of life with less variation, by having the benefit of 'smart' Resources. In this case, 'smart' is an outcome. There isn't less variation because 'smart' Resources somehow have inherently higher GMAT scores; there is less variation, because we did the hard work of reducing it and teaching Resources how to use a process that produced less of it. There remained the question of capacity – the question of whether there was sufficient reserve capacity to handle the remaining variation and uncertainty, whether everything was just a little too buttoned-up, of whether we were counting on drawing to a proverbial inside-straight in order to meet our target.

"We addressed the capacity question in the two versions of Game 4, by going large, by buying the proposition – the mental model – of 'more-for-more'. The additional capacity obtained from having Resources that were strong as well as smart looks like it was worth the additional cost.

"That's where we currently stand", she said.

"By now, playing these games should have given you a sense for how difficult it is to achieve balanced flow and production in a system with balanced capacity. It's both an illusion and a paradox. We have spoken of this issue previously; it is extremely difficult for a production system with balanced capacity to achieve an even rate of production, precisely because its capacity is evenly distributed, with the same capacity at every resource.

"Intuitively, we believe the opposite, that a balanced system produces balanced results", she continued. "But, it's not true, due to variation inherent in the system. In a production system with balanced capacity, variation and uncertainty *anywhere* in the system will affect production *everywhere* in the system, making it impossible to control or predict.

"Moreover, production systems with balanced capacity tend to be very rigid and difficult to manage; they are cumbersome. They are not the adaptable, agile, elegant, easily-managed system that this company needs.

"Yes, we want even-flow production – predictable, even production flow; but, predictable and even production flow is an outcome, not a mechanism. In order to have an even rate of production, we have to purposely unbalance the system that produces it. It requires acknowledging what already exists, what is already true, which is some limitation on what the system can produce.

"We have to acknowledge constraints.

"Constraints exist, whether we want them to exist or not", she said. "If there was no constraint, a production system would theoretically have unlimited capacity. Yet, we know that systems do not have unlimited capacity. There is always a constraint – some internal limit to what a system can produce – and we are better served embracing the constraint and subordinating everything else to it, than we are fighting for a system with balanced capacity. In a system with balanced capacity, the constraint can be anywhere, over time, it will likely appear in more than one where. A system with balanced capacity is not a manageable prescription.

"We have to live with the nature and location of some constraints: An external constraint, like the housing market, or the case of an internal constraint, like a subcontractor that is in high demand or in short supply. Our preference would be to determine the constraint ourselves and place it where we want, where it does the most good, as a pacemaker that pulls work into the system. Recall the discussion on the CCR – the Constraint Capacity Resource.

"And – the acknowledgement of a CCR means that we also need to live with some level of excess/reserve capacity on the non-constraint, non-pacemaker resources.

"All of this, we are going to learn, courtesy of some changes in the game rules", she said. "There are two ways we can reflect constraints and systems with unbalanced capacity.

"One way is to use Resources with different attributes", she explained. "Up to this point, we have employed Resources that were identical, in terms of dice probability and workrate/capacity: four standard Resources, or four 'smart' Resources, or, four 'strong' Resources. However, none of those scenarios resulted in a manageable, predictable system constraint; they just perpetuated the myth of the production system with balanced capacity.

"We can take one Resource, assign it a more-constrained workrate, and make it the Constrained Capacity Resource. It's not a bad approach, although it does have the effect of mandating more expensive Resources, and therefore, incurring higher Operating Expense, with possibly no benefit. The use of more capable, more expensive Resources is what we have had to resort to thus far, in order to overwhelm the effect of variation on a system with balanced capacity."

"What's the other way?", asked the CEO.

"The other way is to let the constraint reflect the result of improving the process, of removing muda – of removing waste and non-value-adding effort – from the workflow, which is quite different than simply reducing variation", replied the intrepid, results-based consultant. "That is really the objective of any process of continuous improvement, be it TQM, Lean, Six Sigma, BPI, its more radical cousin BPR, or Theory of Constraints.

"Six Sigma is arguably more concerned with variation, but every improvement methodology, in some way, deals with getting rid of non-value-adding work and making necessary work flow more evenly, more consistently."

"How would we reflect the value – the benefit – of a better process in one of these games?", asked a sales representative.

The intrepid, results-based consultant pulled up another PowerPoint slide.

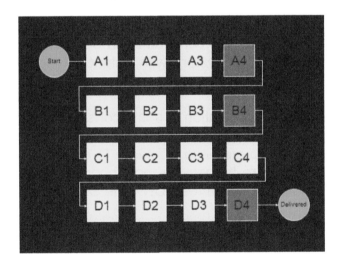

"Here's one way we can do it", she said. "We can show the benefit of an improved process as the A, B, and D phases having fewer tasks to perform. The missing tasks were muda, which no one is going to miss, and now the phase delivers the same value – the same Throughput – with less wasted effort, less duration; the Resource has the same workflow capability it had before the improvement, so now it has reserve capacity.

"The C-Phase didn't register improvement, evidenced by the absence of fewer tasks; perhaps every task in the C-Phase was necessary; perhaps it just couldn't be improved as much, relative to the other three phases. In any event, the C-Phase has no reserve capacity, and it becomes the system's constraint – i.e., its Constrained Capacity Resource, its CCR.

"Because the Throughput of the entire system is determined by the constraint, a production system with unbalanced capacity using four 'smart' Resources has exactly the same projected Throughput and Operating Expense as a production system with balanced capacity using four 'smart' Resources;  and, that would also be true if both systems used 'strong' Resources.

"In either case, the difference is, that an unbalanced system has a Constrained Capacity Resource to focus upon, and a certain amount of excess/reserve capacity on the remaining non-CCRs to support it.

"It's a much more manageable, more predictable production system.  It's a more productive, more profitable production system.  It's a production system that generates a higher Return on Invested Assets.

"Game 5:  Do you want to use 'smart' Resources or 'strong' Resources?"

"Keep the same progression, and run Game 5 with the four 'smart' Resources", said the CEO.  "An improved process with more capable resources.  Good picture of where we want to be."

"Okay, 'smart' Resources it is", she said.  "Before we begin play, let's go over the game rules.  There are four 'smart' Resources – A, B, C, and D – that can roll anything on the dice face, 1-2-3-4-5-6, with a roll of 1 or 2 = 3, a roll of 3 or 4 = 4, and a roll of 5 or 6 = 5;  an average workrate/capacity of 4;  the 3-4-5 profile produces a 33%-33%-33% probability distribution.

"What is the budget?  What are the projections?"

"We have 'smart' Resources, so the anticipated Throughput is 24 completed jobs/projects", said the CFO.  "Operating Expense will absorb 16 jobs/projects;  that being the case, Net Income is anticipated to be eight completed jobs/projects.  Net Income Margin should be 33%.

"Now that we have introduced a system with unbalanced capacity, it's difficult to say how much Inventory we should have.  Let's agree that the combination of more consistent 'smart' Resources with fewer tasks in A, B, and D gives us a better chance than we had in Games 1 and 2 of keeping it at four jobs/projects-in-process, but let's also agree that we won't put a maximum on WIP.  If we average four jobs/projects-in-process throughout the system, Inventory Turn should be 6.0x, and cycle time should be 60 days."

ROIA = .331 X 6.0
ROIA = 1.98
ROIA = 198%

"That should give us an ROIA of almost 200%, so it looks like we're back to square-one", he said, "but with a better chance of pulling it off, because we have smarter Resources and a more manageable production system."

"Play it", said the intrepid, results-based consultant

GAME 5

Game 5 was played using a production system with unbalanced capacity and an improved process; all four Resources were 'smart', which meant they had an average workrate/capacity of four, a roll pattern of 3-4-5, and a 33%-33%-33% probability. Resources A, B, and D each had three tasks to perform in every period/round; only Resource C had four tasks per period/round, and thus became the Constrained Capacity Resource.

Game 5 started with one job/project-in-process in A3, B3, C4, and D3, those tasks being the so-called 'phase-complete' tasks from which the successor Resource could pull work. The level of Inventory/WIP maintained during Game 5 was left to the discretion of the individual Resources, with no stipulated minimum or maximum level of jobs/projects-in-process.

| | G-1 | G-2 | G-3 | G-4 | G-5 | G-6 | G-7 | G-8 |
|---|---|---|---|---|---|---|---|---|
| Description<br>Std. = 4<br>Smart = 4<br>Strong = 5 | Balanced, 4 Std. unlimited WIP | Balanced, 4 Std. Max WIP = 4 | Balanced, 4 Smart WIP = 4 to 6 | Balanced, 4 Strong WIP = 8 Min/Max | Unbal.cap 4 Smart WIP = 4 at Start, unlimited | | | |
| | | | | | | | | |
| **Budgeted** | | | | | | | | |
| T | 24 | 24 | 24 | 30 | 24 | | | |
| I | 4 | 4 | 6-8 | 8 | 4 | | | |
| OE | 16 | 16 | 16 | 20 | 16 | | | |
| P-Rate | 1.5 | 1.5 | 1.5 | 1.5 | 1.5 | | | |
| Inv. Turn | 6.0x | 6.0x | 3.4x | 3.8x | 6.0x | | | |
| CT (days) | 60 | 60 | 90-120 | 96 | 60 | | | |
| NI | 8 | 8 | 8 | 10 | 8 | | | |
| NI % | 33% | 33% | 33% | 33% | 33% | | | |
| ROIA | 200% | 200% | 112% | 125% | 200% | | | |
| | | | | | | | | |
| **Actual** | | | | | | | | |
| T | 19 | 16 | 22 | 29 | 25 | | | |
| I | 8.0 | 4.0 | 7.0 | 8.0 | 9.4 | | | |
| OE | 16 | 16 | 16 | 20 | 16 | | | |
| P-Rate | 1.2 | 0 | 1.4 | 1.5 | 1.6 | | | |
| Inv. Turn | 2.4x | 4.0x | 3.1x | 3.6x | 2.7x | | | |
| CT (days) | 150 | 90 | 115 | 99 | 135 | | | |
| NI | 3 | 0 | 6 | 9 | 9 | | | |
| NI % | 16% | 0% | 27% | 31% | 36% | | | |
| ROIA | 38% | 0% | 84% | 112% | 97% | | | |

| | | | | | | | | |
|---|---|---|---|---|---|---|---|---|
| Resource A | 4.6 | 4.1 | 3.9 | 4.9 | 4.2 | | | |
| Resource B | 3.5 | 4.5 | 3.9 | 5.1 | 4.0 | | | |
| Resource C | 4.0 | 4.1 | 4.2 | 5.0 | 4.1 | | | |
| Resource D | 3.9 | 3.8 | 4.0 | 4.8 | 4.3 | | | |

"Give me the results", said the intrepid, results-based consultant.

"It was interesting", said the CFO. "Game 5 was the first time we exceeded projected Throughput – we completed 25 jobs/projects, one more job/project than the budget of 24. Operating Expense was 16, so our Net Income was nine completed jobs/projects and our Net Income Margin was 36%; above projections, all."

"Here is the interesting part.

"We made those business numbers, despite operating numbers that I thought would have precluded it. Inventory averaged 9.4 jobs/projects-in-process – more than twice the level we projected. As a result, Inventory Turn was only 2.7x, instead of the anticipated 6.0x; cycle time was 135 days – the worst since Game 1, and more than double the 60 day duration that was expected.

"As a result, Return on Invested Assets was 97%, below expectations. ROIA was a composite of the higher-than-expected performance on margin and the lower-than-expected performance on velocity.

"So – we are pleased with our performance, but we are left to wonder how much better that performance could have been."

"The average level of Inventory/WIP during the entire game doesn't tell the entire story. For the front-half of the game – during the first 12 rounds/periods – we averaged 7.8 jobs/projects-in-process. For the back-half of the game, we averaged 11.0 jobs/projects-in-process; during the last quarter of the game – the final six rounds/periods – we averaged a whopping 12.2 jobs/projects-in-process. WIP gradually-and-steadily increased throughout the game.

"All of the Resources performed well", the CFO continued. "There was an 8% variance between the most productive and the least productive Resource; all four Resources were at-or-above the average roll of 4, only 25% of the rolls were below that average, and all four Resources had approximately the same percentage of below-average rolls."

"So – what adjustments would you make?", asked the intrepid, results-based consultant. "What would you change – and what would you not change?"

"The only change I think we should make is regarding Inventory", said the VP of Construction. "By now, we should know better than to start a game with what I would term bare-bones WIP, because having just four jobs/projects-in-process at the beginning of a game is probably less than what we should consider Minimum WIP; there's no buffer for any variation.

"I think we should start higher, and put a limit on the overall level of Inventory in the system. However, Resource C is the Constrained Capacity Resource, and the existence of that constraint is a condition that we have to consider, in the determination of how much Inventory we need."

"What do you mean?", asked a superintendent. "Consider where we place the WIP? Or, determine how much WIP we allow?"

"I want to do both", said the VP of Construction. "Let's impose a Maximum WIP of six jobs/projects-in-process throughout the system. But, let's stipulate that there has to be a minimum of two jobs/projects-in-process in front of Resource C, which is the system's constraint. That should give us adequate protection on the CCR, so that it stays busy. Let's also require that those two jobs/projects-in-process be in the A3 and B3 tasks, meaning they are complete in those phases and ready for the next Resource to pull into its phase."

"And – your expectations?", asked the CEO.

"We are keeping the four 'smart' Resources", replied the CFO. "Throughput should still be 24 completed jobs/projects, Operating Expense will consume 16 completed jobs/projects, Net Income should be eight jobs/projects, and Net Income Margin should be 33%.

"Worse case, we're at Max WIP every round/period, so Inventory Turn is 4.0x, cycle time is 90 days. ROIA should be 132%."

"Let's play Game 6", said the intrepid, results-based consultant.

GAME 6

Game 6 is an improved production system with unbalanced capacity, four 'smart' Resources, average workrate of four, 3-4-5 pattern, 33%-33%-33% probability. A, B, and D have three tasks each; C has four tasks.

The game started with one job/project-in-process each in the A3, B3, C4, and D3 'phase-complete' tasks from which the successor Resource could pull work. Max WIP was six jobs/projects-in-process; two jobs/projects-in-process always in front of C, preferably in A3 and B3.

"How did your changes work?", asked the intrepid, results-based consultant.

"Game 6 was basically the same as Game 5", said the CFO, "just the changes to the amount and location of jobs/projects-in-process. Because we had required more Inventory, we projected a lower Inventory Turn, longer cycle times, and, therefore, a lower Return on Invested Assets, despite expecting Throughput, OE, and Net Income to be unchanged.

|  | G-1 | G-2 | G-3 | G-4 | G-5 | G-6 | G-7 | G-8 |
|---|---|---|---|---|---|---|---|---|
| Description<br>Std. = 4<br>Smart = 4<br>Strong = 5 | Balanced, 4 Std. unlimited WIP | Balanced, 4 Std. Max WIP = 4 | Balanced, 4 Smart WIP = 4 to 6 | Balanced, 4 Strong WIP = 8 Min/Max | unbal.cap 4 Smart WIP = 4 at Start, unlimited | unbal.cap 4 Smart Max WIP = 6, 2 in A/B | | |
| | | | | | | | | |
| **Budgeted** | | | | | | | | |
| T | 24 | 24 | 24 | 30 | 24 | 24 | | |
| I | 4 | 4 | 6-8 | 8 | 4 | 6 | | |
| OE | 16 | 16 | 16 | 20 | 16 | 16 | | |
| P-Rate | 1.5 | 1.5 | 1.5 | 1.5 | 1.5 | 1.5 | | |
| Inv. Turn | 6.0x | 6.0x | 3.4x | 3.8x | 6.0x | 4.0x | | |
| CT (days) | 60 | 60 | 90-120 | 96 | 60 | 90 | | |
| NI | 8 | 8 | 8 | 10 | 8 | 8 | | |
| NI % | 33% | 33% | 33% | 33% | 33% | 33% | | |
| ROIA | 200% | 200% | 112% | 125% | 200% | 132% | | |
| | | | | | | | | |
| **Actual** | | | | | | | | |
| T | 19 | 16 | 22 | 29 | 25 | 23 | | |
| I | 8.0 | 4.0 | 7.0 | 8.0 | 9.4 | 5.0 | | |
| OE | 16 | 16 | 16 | 20 | 16 | 16 | | |
| P-Rate | 1.2 | 0 | 1.4 | 1.5 | 1.6 | 1.4 | | |
| Inv. Turn | 2.4x | 4.0x | 3.1x | 3.6x | 2.7x | 4.6x | | |
| CT (days) | 150 | 90 | 115 | 99 | 135 | 78 | | |
| NI | 3 | 0 | 6 | 9 | 9 | 7 | | |
| NI % | 16% | 0% | 27% | 31% | 36% | 30% | | |
| ROIA | 38% | 0% | 84% | 112% | 97% | 140% | | |
| | | | | | | | | |
| Resource A | 4.6 | 4.1 | 3.9 | 4.9 | 4.2 | 4.4 | | |
| Resource B | 3.5 | 4.5 | 3.9 | 5.1 | 4.0 | 4.1 | | |
| Resource C | 4.0 | 4.1 | 4.2 | 5.0 | 4.1 | 4.0 | | |
| Resource D | 3.9 | 3.8 | 4.0 | 4.8 | 4.3 | 4.0 | | |
| | | | | | | | | |

"We were one completed job/project short on Throughput – 23 completed jobs/projects instead of 24 – which also reduced Net Income by one job/project. As a result, we had a slightly lower Net Margin – 31% instead of 33% – on slightly less Throughput.

"However, we also had less average WIP than we had anticipated, averaging five jobs/projects-in-process, instead of the six we had expected. As a result, we had a better-than-projected Inventory Turn – 4.6x instead of 4.0x – and a shorter cycle time, 78 days instead of 90 days; the best news, Return on Invested Assets was 140%, ahead of the 132% that was consensus.

"We would have done even better, if our Resources had cooperated", said the CFO. "Resource C – the CCR – had an average workflow rate of 4.0, meaning, the most-constrained Resource had the lowest average roll; Resources A, B, and C all averaged a roll of 4.0 or higher – but, they aren't the Capacity Constraint Resource. If we had rolled anything better on Resource C, our Throughput would have likely been above projections."

"What does all that tell us?", asked the intrepid, results-based consultant.

"It tells me that we are getting faster", said the CEO. "And – that we are getting better at managing production systems with unbalanced capacity, getting better at Inventory decisions, and getting better at exploiting the constraint. Our Inventory actually declined as the game progressed.

"We are mindful of how you taught us to look at the metrics of a pipeline – the metrics of a production system. How a pipeline's *size* is defined by the amount of work-in-process it is intended – or designed – to carry", said the CEO. "How a pipeline's *length* is its cycle time. How a pipeline's *capacity* is defined as the rate of output – or Throughput – a pipeline that size can produce, with a planned, finite, and controlled level of work-in-process."

"You have good recall", said the intrepid, results-based consultant. "That was an interesting term you used – 'exploiting the constraint'.

"What did you mean?"

"Granted, constraints that are internal to the system can be policies and such", explained the CEO. "However, when the constraint is the capacity of an internal resource, that resource – the constraint resource – is what determines the throughput, or output, of the entire system, because it is either the resource with the least capacity, or is the resource in greatest demand.

"By definition, every other resource has either more capacity or less demand than the constraint, therefore, every other resource has excess or reserve capacity.

"The non-constraint resources have to support the work of the constraint; all of the non-constraint resources upstream of the constraint have to work together to make certain that the constraint resource is never idle, to make certain that it works all the time, mainly by using their excess/reserve capacity to make certain the constraint resource always has enough to work on.

"That's what I mean by 'exploiting the constraint': Getting as much output as you possibly can from it. Throughput lost on the constraint is Throughput lost to the system, and

Throughput lost to the system is Throughput lost forever. Time doesn't stand still, we can't make more of it.

"That's the TOC description – the constraint-management description. As you explained in a previous session, you would get, more-or-less, the same explanation from Lean Thinking; in a build-to-order process, like homebuilding, Lean would make the constraint resource its pacemaker resource – same difference. In either case, it would enable us to pull new jobs/projects into the system based on the performance of the constraint/pacemaker."

"Well stated", said the intrepid, results-based consultant.

"And, close-enough; that's what we well-state to our framers. Can we agree that Game 6's performance is as good as it gets?", asked a superintendent.

"No", said the intrepid, results-based consultant. "Question: What can we do to make it better?"

She waited. Finally, she sighed, looked at the CFO, and said, "C'mon. You just said it: 'If we had rolled anything better on Resource C, our Throughput would have likely been above projections'." She pointed to the last PowerPoint slide that she had pulled up, still sitting on the screen.

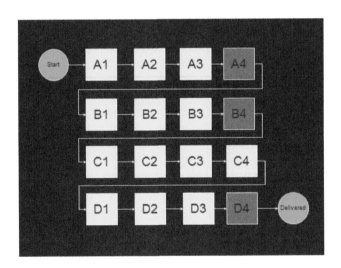

"What is the potential problem with Resource C in this purposely-unbalanced system?", she asked. "Resource C is the constraint resource. That's not the problem. We put it there. What's the problem?"

"The problem is: How do we elevate the constraint?", the senior superintendent said. "Having done everything we need to do – not everything we can do, but everything we need to do – to exploit the constraint, we have no choice, except to now elevate the constraint; the only action we can take is to give the constraint resource more capacity; and the only way we can give the constraint resource more capacity is to substitute a 'strong' Resource for a 'smart' Resource at the C-Phase.

156

"We don't take that decision lightly", he said. "Making this change will increase Operating Expense – i.e., increase the fixed, non-variable cost associated with Resource C. That decision will increase our Operating Expense from 16 jobs/projects per game, to 17 jobs/projects per game.

"We reduced variation in the system, by switching-out all standard Resources with all 'smart' Resources. We tried a 'more-for-more' approach across the board, by substituting all 'strong' Resources for 'smart' Resources, which didn't work out so well; it made us bigger, not necessarily more profitable.

"We adjusted and controlled the level of Inventory.

"We improved the process, by removing muda and non-value-adding work, which reduced tasks in three of the four phases; and in doing so, we changed a production system with balanced capacity into a more manageable production system with unbalanced capacity.

"As a 'smart' Resource, Resource C – the constraint resource – has a workrate that matches the amount of work required to complete the C-Phase; average roll of 4, four tasks to complete. As we've seen, however, a 'smart' Resource also has the possibility of underperforming, of rolling a 3. I think we should acquire a 'strong' Resource as Resource C. A 'strong' Resource is already a 'smart' Resource; a 'strong' Resource and a 'smart' Resource have the same dice roll probability, the same variance profile.

"But – a 'strong' Resource has a profile that will always allow it to complete at least one job/project per round; sometimes, it will be able to complete more than one job/project per period. Substituting a 'strong' Resource for a 'smart' Resource in C-Phase reduces the amount of excess/reserve capacity on A, B, and D – the non-constraint resources – relative to C, but not enough to rebalance the system or to change/remove – in effect, blur – the constraint."

"How do you know?", asked the intrepid, results-based consultant.

"You can check my numbers on your coelacanth, but the reserve ratio is still higher on Resource C than it is on A, B, or D", he replied. "In Games 5 and 6, the non-constraints (A, B, and D) had a ratio of 1.33:1, the result of having an average roll of 4, but only having three tasks to perform. In those two games, the CCR (Resource C) had a 1:1 ratio – average roll of 4 and four tasks; Resource C was a 'smart' Resource, but it had no reserve capacity.

"In this game – Game 7 – Resource C will be a 'strong' Resource with a reserve ratio of 1.25:1, based on an average roll of 5 to deliver four tasks. That's still less than A, B, and D with a ratio of 1.33:1. So – we might find that we have to adjust the difference between ratios, but Resource C will remain the CCR, even with some needed protective/reserve capacity.

"There was evidence in Game 6 that, at times, we may have starved Resource C, the constraint resource. Acquiring a 'strong' Resource for the C-Phase will increase the average workrate/capacity from four to five, and make it even more likely that Resource A and

Resource B will struggle to maintain a sufficient number of jobs/projects-in-process. I think there is an argument for maintaining more Inventory in front of Resource C.

"But – this should work", the senior superintendent said. "It will elevate the constraint."

"Give me the projections", said the CEO.

"We are now going to use three 'smart' Resources and one 'strong' Resource", said the CFO. "As a result, Throughput should be 30 completed jobs/projects, Operating Expense will consume 17 completed jobs/projects, Net Income should be 13 jobs/projects, and Net Margin should be 43%.

"We averaged five jobs/projects-in-process during Game 6, so let's say we increase it slightly, like Senor' here recommends, and say that we expect to average six jobs/projects-in-process for Game 7. That would give us an Inventory Turn of 5.0x, and a resulting cycle time of 72 days."

The CFO walked to the erasable board and made the appropriate changes.

$$ROIA = (Net\ Income \div Throughput) \times (Throughput \div Inventory)$$
$$ROIA = Return\ on\ Sales \times Inventory\ Turn$$
$$ROIA = Margin \times Velocity$$

$$ROIA = (13 \div 30) \times (30 \div 6)$$
$$ROIA = .43 \times 5.0$$
$$ROIA = 2.15$$
$$ROIA = 215\%$$

"ROIA should be 215%", said the CFO. "I'll take my bonus right now."

"Without us, you wouldn't have anything", said the VP of Sales. "We make it rain new home sales."

"Keep it coming, Rainmaker", said one of the superintendents, who had also been the Resource C since Game 4. "The F-150 and Silverado crowd appreciates it."

The CFO deadpanned the superintendent. "Do you recall the very first session, and we were talking about variable costs versus non-variable costs, who was what? You asked: 'What about us? What are we?'. At the time, I said you were a totally worthless, soon-to-be-eliminated-thus-no-longer-non-variable cost. Remember?"

The CFO picked up the superintendent's dice, handed it to one of the sales representatives, and said: "You're our go-to girl."

"I agree", said the CEO.

"This will be Game 7, everyone", said the intrepid, results-based consultant.

GAME 7

Game 7 was played using a production system with unbalanced capacity, an improved process, and a constraint resource.

Resources A, B, and D – three of the four Resources – were 'smart' Resources, which meant they had an average workrate/capacity of four, a roll pattern of 3-4-5, and a 33%-33%-33% probability.  Resources A, B, and D also held in-common the fact that they only had three tasks to perform in their respective A, B, and D Phases during every period/round, instead of four;  the reduced task load on these Resources A, B, and C was the product of an improved process, a process with less waste, a process with less non-value-adding work.

Resource C was different than the other three Resources.  Resource C was a 'strong' Resource, which meant that it had the same roll probability as a 'smart' Resource – 33%-33%-33% – but, had a roll pattern of 4-5-6, and, therefore, an average workrate/capacity five.  Resource C had four tasks per period/round in its phase, making it the Constrained Capacity Resource.

Game 7 started with five jobs/projects-in-process:  one job/project-in-process in A3, one in B2 and B3, one in C4, and one in D3.  The stipulated Inventory for G-7 was a minimum of three jobs/projects-in-process in Resources A and B, and a maximum of six jobs/projects-in-process in the system.

| | G-1 | G-2 | G-3 | G-4 | G-5 | G-6 | G-7 | G-8 |
|---|---|---|---|---|---|---|---|---|
| Description<br>Std. = 4<br>Smart = 4<br>Strong = 5 | Balanced, 4 Std. unlimited WIP | Balanced, 4 Std. Max WIP = 4 | Balanced, 4 Smart WIP = 4 to 6 | Balanced, 4 Strong WIP = 8 Min/Max | Unbal.cap 4 Smart WIP = 4 at Start, unlimited | Unbal.cap 4 Smart Max WIP = 6, 2 in A/B | Unbal.cap 3 Smart 1 Strong Max WIP = 6, 3 in A/B | |
| | | | | | | | | |
| **Budgeted** | | | | | | | | |
| T | 24 | 24 | 24 | 30 | 24 | 24 | 30 | |
| I | 4 | 4 | 6-8 | 8 | 4 | 6 | 6 | |
| OE | 16 | 16 | 16 | 20 | 16 | 16 | 17 | |
| P-Rate | 1.5 | 1.5 | 1.5 | 1.5 | 1.5 | 1.5 | 1.8 | |
| Inv. Turn | 6.0x | 6.0x | 3.4x | 3.8x | 6.0x | 4.0x | 5.0x | |
| CT (days) | 60 | 60 | 90-120 | 96 | 60 | 90 | 72 | |
| NI | 8 | 8 | 8 | 10 | 8 | 8 | 13 | |
| NI % | 33% | 33% | 33% | 33% | 33% | 33% | 43% | |
| ROIA | 200% | 200% | 112% | 125% | 200% | 132% | 215% | |
| | | | | | | | | |
| **Actual** | | | | | | | | |
| T | 19 | 16 | 22 | 29 | 25 | 23 | 27 | |
| I | 8.0 | 4.0 | 7.0 | 8.0 | 9.4 | 5.0 | 5.9 | |
| OE | 16 | 16 | 16 | 20 | 16 | 16 | 17 | |
| P-Rate | 1.2 | 0 | 1.4 | 1.5 | 1.6 | 1.4 | 1.6 | |
| Inv. Turn | 2.4x | 4.0x | 3.1x | 3.6x | 2.7x | 4.6x | 4.6x | |
| CT (days) | 150 | 90 | 115 | 99 | 135 | 78 | 79 | |
| NI | 3 | 0 | 6 | 9 | 9 | 7 | 10 | |
| NI % | 16% | 0% | 27% | 31% | 36% | 30% | 37% | |
| ROIA | 38% | 0% | 84% | 112% | 97% | 140% | 170% | |
| | | | | | | | | |
| Resource A | 4.6 | 4.1 | 3.9 | 4.9 | 4.2 | 4.4 | 3.7 | |
| Resource B | 3.5 | 4.5 | 3.9 | 5.1 | 4.0 | 4.1 | 3.9 | |
| Resource C | 4.0 | 4.1 | 4.2 | 5.0 | 4.1 | 4.0 | 5.2 | |
| Resource D | 3.9 | 3.8 | 4.0 | 4.8 | 4.3 | 4.0 | 3.9 | |
| | | | | | | | | |

"How did we do?", asked the intrepid, results-based consultant.

"Except for the small matter of managing expectations", the CFO said, "we did quite well. We eclipsed the performance we achieved in any previous game; at 37%, Game 7 produced our highest-ever Net Income Margin, and, at 4.6x, tied our highest-ever Inventory Turn.

"That performance on the two components of Return on Invested Assets pushed that ultimate measure of economic return to 170%, more than 20% higher than our previous high score.

"Throughput was lower than expected – 27 completed jobs/projects versus the 30 completed jobs/projects that were anticipated. Inventory was almost exactly where we projected, a good indicator of the operational discipline we are developing. Operating Expense was where we expected it to be, so the residual Net Income was lower – 10 completed jobs/projects versus the 13 jobs/projects we had projected.

"As good as this performance was, overall, it did not meet expectations; that's nothing new, we don't ever achieve what we have projected", the CFO said. "Net Margin was 37%, but we had projected 43%. Return on Invested Assets was 170%; we had anticipated that ROIA would be 215%. Turn was 4.6x; we expected 5.0x; and, so on."

"This strikes me as being a function of how Resources performed in this specific game", said the CEO. "Resource C – the Capacity Constraint Resource – was the only Resource whose performance exceeded its average workflow/capacity; it had an average roll of 5.2 tasks per period. Resources A, B, and D – the non-constraint resources that need to support the constraint resource – didn't achieve their average; they averaged rolls of 3.7, 3.9, and 3.9, respectively.

"It's pretty simple: It doesn't matter how much capacity we give the constraint resource, if the upstream non-constraint resources – Resources A and B – don't provide enough jobs/projects-in-process to keep Resource C from being idle. It doesn't matter how much work Resource C performs, if the downstream non-constraint resource – Resource D – doesn't keep pace.

"Consider this: 42% of the combined periodic work performed by the three non-constraint resources was below their average workflow/capacity; statistically, it should have been 33%. The constraint resource – Resource C – is a 'strong' Resource that will never produce less than the amount of work required to complete at least one job/project per period; nevertheless, just 21% of its periodic work was below its average workflow/capacity.

"Resource C outperformed all of the other Resources", he said, nodding to the sales representative that had been drafted as the replacement Resource C for Game 7. "Our go-to girl simply outworked you boys."

"It was hard work for me and my sales team", said the VP of Sales. "But, you are so incredibly fortunate that we personally arranged for that limitless supply of ready-to-start jobs/projects, sitting in the Start circle, so Go-To Girl and the Guys could start as many jobs/projects as they desired."

"Keep makin' it rain, Chief", said the CEO.

"We shall see", said the intrepid, results-based consultant. "Run Game 7 again, as Game 8."

## GAME 8

Game 8 had a production system with unbalanced capacity, benefiting from an improved process in three phases, with a 'strong' constraint resource supported by 'smart' non-constraint resources.

"Tell us about the results of Game 8, and contrast it with the results of Game 7", said the intrepid, results-based consultant, "because these were two games with identical rules, achieving different results."

"In some ways, the difference between Game 7 and Game 8 was a continuing improvement in our ability to 'see' production more clearly as a system", said the CFO. "It was also the law of averages, because Resources A, B, and D clearly performed closer to expectations.

"On the second point, in Game 7, the three non-constraint resources performed below average 42% of the time; in Game 8, it occurred only 22% of the time, so they were able to support the system's constraint resource – Resource C – more effectively. Resources A, B, and D also produced more work; instead of respective average rolls of 3.7, 3.9, and 3.9, they had respective average rolls of 3.9, 4.1, and 4.3.

"On the first point, in Game 7, Resources A and B could not compensate for each other's insufficient production; in Game 8, Resource B's higher production could more readily offset Resource A's lower production. Resource A and Resource B, working together, could adequately support Resource C with the amount of Inventory between them; they chose to keep a higher protective level of Inventory, which was the right call."

"The results speak for themselves", said the CEO. "But, I'll speak anyway; from the perspective of any measure of operating performance or business outcome – from the matter of Throughput, Inventory, Operating Expense, from the standpoint of productivity rate, Inventory Turn, cycle time, from the basis of Net Income, Net Margin, or Return on Invested Assets – Games 7 and 8 were the best outcomes that we produced."

| Description<br>Std. = 4<br>Smart = 4<br>Strong = 5 | G-1<br>Balanced, 4 Std. unlimited WIP | G-2<br>Balanced, 4 Std. Max WIP = 4 | G-3<br>Balanced, 4 Smart WIP = 4 to 6 | G-4<br>Balanced, 4 Strong WIP = 8 Min/Max | G-5<br>Unbal.cap 4 Smart WIP = 4 at Start, unlimited | G-6<br>Unbal.cap 4 Smart Max WIP = 6, 2 in A/B | G-7<br>Unbal.cap 3 Smart 1 Strong Max WIP = 6, 3 in A/B | G-8<br>Unbal.cap 3 Smart 1 Strong Max WIP = 6, 3 in A/B |
|---|---|---|---|---|---|---|---|---|
| **Budgeted** | | | | | | | | |
| T | 24 | 24 | 24 | 30 | 24 | 24 | 30 | 30 |
| I | 4 | 4 | 6-8 | 8 | 4 | 6 | 6 | 6 |
| OE | 16 | 16 | 16 | 20 | 16 | 16 | 17 | 17 |
| P-Rate | 1.5 | 1.5 | 1.5 | 1.5 | 1.5 | 1.5 | 1.8 | 1.8 |
| Inv. Turn | 6.0x | 6.0x | 3.4x | 3.8x | 6.0x | 4.0x | 5.0x | 5.0x |
| CT (days) | 60 | 60 | 90-120 | 96 | 60 | 90 | 72 | 72 |
| NI | 8 | 8 | 8 | 10 | 8 | 8 | 13 | 13 |
| NI % | 33% | 33% | 33% | 33% | 33% | 33% | 43% | 43% |
| ROIA | 200% | 200% | 112% | 125% | 200% | 132% | 215% | 215% |
| | | | | | | | | |
| **Actual** | | | | | | | | |
| T | 19 | 16 | 22 | 29 | 25 | 23 | 27 | 30 |
| I | 8.0 | 4.0 | 7.0 | 8.0 | 9.4 | 5.0 | 5.9 | 6.5 |
| OE | 16 | 16 | 16 | 20 | 16 | 16 | 17 | 17 |
| P-Rate | 1.2 | 0 | 1.4 | 1.5 | 1.6 | 1.4 | 1.6 | 1.8 |
| Inv. Turn | 2.4x | 4.0x | 3.1x | 3.6x | 2.7x | 4.6x | 4.6x | 4.6x |
| CT (days) | 150 | 90 | 115 | 99 | 135 | 78 | 79 | 78 |
| NI | 3 | 0 | 6 | 9 | 9 | 7 | 10 | 13 |
| NI % | 16% | 0% | 27% | 31% | 36% | 30% | 37% | 43% |
| ROIA | 38% | 0% | 84% | 112% | 97% | 140% | 170% | 198% |
| | | | | | | | | |
| Resource A | 4.6 | 4.1 | 3.9 | 4.9 | 4.2 | 4.4 | 3.7 | 3.9 |
| Resource B | 3.5 | 4.5 | 3.9 | 5.1 | 4.0 | 4.1 | 3.9 | 4.1 |
| Resource C | 4.0 | 4.1 | 4.2 | 5.0 | 4.1 | 4.0 | 5.2 | 5.0 |
| Resource D | 3.9 | 3.8 | 4.0 | 4.8 | 4.3 | 4.0 | 3.9 | 4.3 |
| | | | | | | | | |

"So – what did you learn?", asked the intrepid, results-based consultant, walking to the flipchart. "More importantly – how will you apply those lessons in terms of how you manage RB Builders' production system? These are all production principles we have discussed in previous sessions.

Systems-thinking

Connection – Throughput, Inventory, Operating Expense; T-I-OE

Constraints v. the unlimited capacity myth

V-A-R-I-A-T-I-O-N

Balanced capacity v. Unbalanced Capacity
Resources – non-constraint and constraint
Resources – "smart" v. "strong"
Reserve Capacity and Inventory – buffering the system from variation
Improving the process – removing muda and non-value-adding work

"We aren't going to discuss them right now. When you remember the lessons from the game, remember the principles."

Following the first 'discovery learning' session, RB Builders began involving more teammates in game situations designed to simulate homebuilding production.

RB Builders incorporated new versions of the game, such as: (1) having game scenarios in which multiple building companies competed in a single market for a finite number of available new home sales, in order to demonstrate a market/external constraint, as opposed to an internal constraint; (2) having game scenarios in which prices and margins were determined by a bidding process that reflected the relationship between the potential supply and the available demand for new homes in a housing market: and, (3) having game situations in which finite resources had to be deployed across multiple new home communities with different volume, prices, and margins.

Playing the game continued to be about creating an intuitive, instinctive feel for production principles that readily transferred into analyzing conditions encountered in the field.

The game was always about The Pipeline, about the Connection, and about systems-thinking. It was always about how a production system manages process variation and uncertainty, how a production system is scheduled, and how production is managed as a project portfolio.

# CHAPTER VIII: CONTEXT AND CONTENT

"Congratulations", said the intrepid, results-based consultant.

"You are now bonafide experts, collectively and individually, in the principles and disciplines that govern homebuilding production. That distinction may not mean much for awhile, considering the current state of the housing market[12] and condition of the homebuilding industry.

"At some point, it will surely mean a great deal.

"There are any number of requirements and necessary conditions that must be met, in order to live above your competition – that is, for RB Builders to create and sustain a degree of competitive separation. For certain, you need compelling product, good community locations, all of that. But – the hardest attribute for any competitor to match is velocity.

"Speed, productivity, and agility are killer attributes.

"When the decision was made to build an enterprise-wide understanding of production physics, it was not the only initiative on your plate. It exists within the context of the results-focused commitment to improve RB Builders' operating performance and economic return that we are now engaged in. The owners of this company are constantly and consistently emphasizing the importance of achieving targeted results. They are now sharing the financial performance and condition of the company openly with the entire team, and teaching every team member what is needed to run a homebuilding company as a profitable, sustainable business.

"Over the course of the next two years, we will be involved in a succession of projects with short timeframes and targeted results, each the logical successive step in the pursuit of an overall goal. Some projects will address margins, while others will target productivity. Some will target scheduling, some will target workflow, and some will target product design. Most of the projects will be done in sequence, not in parallel. But – the active projects will always focus on the constraint; it will always focus on improving the performance of whatever it is that limits or restricts RB Builders' ability to generate higher levels of Gross Income.

"That's the way it will go, one project after the next, a true process of continuous improvement. It has, in fact, already started.

"Along the road, RB Builders will need to embrace flexibility and innovation. You will need to experiment with different techniques and approaches, even if you initially shun major systems or technology conversions, out of concerns about your capability to implement the changes.

---

[12] As noted elsewhere, *The Pipeline* is placed at the end of 2007; it was actually written over a five-year period (2007-2012).

"You will need to learn from the outcomes of every project, and incorporate what you learn into each successive project", she said. "As you do more for yourselves, your confidence in your own capacity and capability to manage change and get results will grow. As your capability and your capacity increase, your dependence on me will lessen. And, at some point, my role – certainly as a consultant – will come to an end. In the meantime, it's my job to keep you focused on the outcome.

"Your mental models have begun to change. You are learning how and when to use all of the tools in the toolbox, without regard to the consulting religion or denomination from which they come. You are learning to challenge assumptions and decisions.

"You are learning how to 'see' production, how to look at capacity, productivity, and production planning and management. You have learned what it means to be in the 'business' of homebuilding, not simply what it means to be in the homebuilding business. You are learning how to solve core problems, learning how to resolve conflict, learning how to design new processes, new production systems, and new measurement systems.

"You are learning how to gauge the choices of where and when to expend effort and resources."

She turned to the erasable board and made a final list.

The Pipeline
The Connection
Systems-thinking
Production Systems
Processes
Project Portfolios

"Since we have covered so much, let's summarize, very quickly, everything we have learned about the production principles and disciplines that apply to the production-oriented nature of homebuilding", she said.

"When we talk about production principles and disciplines, we mean (1) how RB Builders visualizes production; (2) how it connects its production planning and management to operating performance and business (financial) outcomes; (3) how it understands production as a system of interdependent, connected parts; (4) how it focuses on solving problems and managing constraints; (5) how it handles variation and uncertainty; and (6) how it manages production from a process, value stream, and project portfolio perspective."

The intrepid, results-based consultant summarized one point at a time.

The Pipeline

"To understand homebuilding production, it helps to have a picture – it helps to have a visual image. The picture of RB Builders' production system is a pipeline. That pipeline has dimensions and measures – a size, a capacity, a length, and a cost – and it has *valves*, which are the mechanisms that control the flow of work that goes through it.

"You have to understand how to utilize it."

## The Connection

"You must connect the physics that control production to the production-related decisions you make every day;  you must connect the decisions you make to the operating performance that drives desired business outcomes.  The key to connecting operating performance to economic outcome is an understanding of the three terms that describe 'what happens to money' in a homebuilding operation:

"Throughput.  Inventory.  Operating Expense."

## Systems-thinking

"We live in a world of systems, and we must learn to think and act in those terms.  The homebuilding industry, the housing and real estate market, and the local and national economies in which a homebuilding company operates – they are all interdependent parts of a system.  The business environment in which a homebuilding company must operate is a system.

"A homebuilding company is not a confederation, not some loosely-connected set of independent, non-dependent, unrelated parts, some random collection of processes, departments, systems, resources, policies, and other isolated pieces of a whole.

"It is a *system* – an integrated arrangement of parts that operate in dependency of each other, all of which must work together to reach a stated, common goal.  Inter-dependent, not independent.

"Systems-thinking is critical to an understanding of production principles and disciplines.  It is a way of thinking, a way of reasoning, a way of solving problems, a way of acting upon information."

## Processes

"You must understand the two core elements involved in managing a production system as a process:  (1) how a production system manages variation and uncertainty – which is the same for a process, a value stream, or a project network – and, (2) how a production system is scheduled."

## Project Portfolios

"You must understand the four main categories of the principles and disciplines for managing production from a project portfolio perspective, from the perspective of multiple houses/jobs being managed simultaneously through a production system: (1) Duration; (2) Structure; (3) Scheduling; and (4) Management."

The intrepid, results-based consultant started a new list on one of the flipcharts.

*RB Builders Integrated Production System (RB-IPS)*

"Everything we have done so far has been about developing an intuitive and instinctive understanding of the principles and disciplines of production", she said. "These principles are like the laws of physics, and if we wanted to state them generally enough, they would apply to a broad range of industry verticals. We have dealt with these production principles and disciplines in the specific context of a homebuilding enterprise. However, whether you consider them generally or specifically, production principles and disciplines – by definition – must act upon something.

"They are like gravity; gravity acts – it has an effect – upon mass.

"The principles and disciplines that govern production are no different. In the case of a homebuilding enterprise, production principles and disciplines act upon its production management system. In the specific case of RB Builders, they act upon what we call the RB Builders Integrated Production System – the RB-IPS.

"The RB-IPS was developed several years ago. It evolved from the production management system that preceded it, what was known as the RB Builders Even-Flow Production System, or RB-EFPS. There was a sustained push for a couple of years – workshops, training, etc., but then came the end of The Age of Homebuilder Entitlement; it's not the only casualty of this housing recession.

"As a production management system, the RB-IPS incorporates many of the principles and disciplines we have been learning. Because it preceded the codification of the production principles and disciplines, there are also now some missing pieces. The RB-IPS nevertheless remains a comprehensive, unified approach to production planning and management, one that creates a deep, visual, intuitive, instinctive understanding of homebuilding production, and provides a unique, proprietary expression of homebuilding production 'branded' to RB Builders.

"The RB-IPS is the production management system; the RB-IPS is not to be confused with the operating/information system, which integrates purchasing, job costing, scheduling, and accounting.

"The RB-IPS has three elements", she said, continuing her list.

*RB Builders Integrated Production System (RB-IPS)*
1. *Daily Jobsite Management and Quality Assurance*
2. *Production Planning and Management*
3. *Epic Partnering*

"It is a production management system in which the day-to-day management of jobs is based on routines, on clear, visual standards, and on intense, rigorous inspection;  one that shares an integrated system of handling all of the scheduling, approvals, and other communication – connectedly, wirelessly, paperless, on-demand, as part of a commitment to streamlining the flow of information and communication across the RB Builders spectrum;  one that supports a method of problem-solving that quickly gets to the root causes of jobsite and construction issues, and resolves them.

"It is a production management system that allows RB Builders to plan and manage production throughout the life of every community, one that gets each community through its three life-stages start-up, sustained production, and exit – through takeoff, level-flight, and final approach;  it is a production management system that delivers on the RB Builders objective of a pipeline generating a maximum rate of closings with a planned, controlled, and finite level of work-in-process;  it reflects RB Builders' understanding that even-flow is an outcome, not a mechanism.

"It is a production management system that specifies the means by which RB Builders fosters epic relationships of mutual interest with its building partners and supply partners", she said. "The RB-IPS provides both the process and the program for progressively transforming subcontractors and suppliers into true partners, into trusted allies, joined together by their shared, mutual interests."

The intrepid, results-based consultant shut down her notebook, and began to pack her materials.  As she did, she addressed the team.

"I want you to know how impressed – and gratified – I am by your efforts, your attention, your attitudes, your cooperation, and your diligence.  Learning new stuff is hard, and this was a lot of new stuff.

"We have much more to do."

She slung her computer bag over her shoulder.  She paused, and then turned to the team and said, "Let me leave you with some final thoughts.

"We have talked about elegant solutions – about clearing distractions, focusing the effort and driving toward the root causes of problems", she said.  "However, there is a chemistry to achieving and sustaining improvements in business performance, and that chemistry implies a certain complexity and a comprehensiveness.

"And, for that reason, improving operating performance and business outcomes is hard, involved work.

"It is *hard* work because neither operating performance nor business outcomes can be improved without doing things differently, and change is disruptive and threatening to people.

"It is *involved* work because improving operating performance and business outcomes requires more than a simple, one-dimensional approach.  Not to the point of causing you to lose focus on the current reality – on the rapid-results project at hand – but, more about the insufficiency of becoming one-dimensional over time. Over a longer period of time, improving operating performance and business outcomes requires continuous effort on different fronts.

"These sessions have been devoted to production principles and disciplines. In my view, improving the velocity side of the ROIA equation is the best path – perhaps the only path – to sustainable competitive separation, but there is more to business performance than managing production."

She returned to the erasable board, and made a short, numbered list.

1. Discipline
2. Context
3. Perspective

"Improving business performance boils down to getting the job done – viewing the issue, sustaining the effort, and getting the results – in three critical dimensions", she said, completing each item.

1. Discipline = narrow the focus, deliver distinctive value

"First, you have to continue to narrow your focus. You cannot be all things to all people. You have to make certain that RB Builders' operating model – its structure, its systems, its processes, who it selects and develops as teammates, what its culture is – delivers exceptional levels of the specific and distinctive value demanded by a narrowly-defined segment of homebuyers.

"It is the choice of a specific value proposition over any other, and how you choose to fulfill it."

2. Context = a company of business-people

"Second, you have to continue to turn RB Builders into a company of business-people, by teaching teammates the real numbers of the business, giving them the authority – and responsibility – to act on that knowledge, and then giving them a real financial stake in the outcome.

"That's what being a 'savvy, accountable, motivated homebuilding team' really means."

3. Perspective = get horizontal

"Third, you need to wrap your minds around the natural, horizontal, process-centered manner in which RB Builders performs work and creates value on behalf of its homebuyers and other stakeholders", she continued. "It is the most basic, the most universal proposition in all of business: The reason an enterprise exists, the way it makes money – is through the value that enterprise delivers to customers and other stakeholders; that value is only delivered by the work that the enterprise performs; that work has to be performed in some method of workflow; and those methods of workflow exist, whether the enterprise is intentional about them or not.

170

"We used to state the sequence of our proposition underlying business process improvement as *value-work-process*. Homebuilding involves both process management and project management, but it's still about workflow, and about the work performed in either process or project management systems.

"Focus on these three dimensions", she said. "Remember them: *Discipline:* Narrow the focus, deliver distinctive value. *Context:* Knowledge, transparency, a stake in the outcome. *Perspective:* Everything in the business model oriented to the requirements of the workflow that creates the value.

"Everything else – the strategies, tactics, policies, scorecards, products, markets, vendors, departments, positions, including everything we have talked about concerning production principles – are means to an end, the side issues to getting the job done in these three dimensions.

"Final points", said the intrepid, results-based consultant. "The words of three men, advice on dealing with dire circumstances.

"When he was President, Ronald Reagan would remind people to open their eyes, and in his words, 'Don't be afraid to see what you see'. It was a phrase he used in his farewell speech; it was offered in the context of dealing with former enemies, circumstances that were less-dire at that point, if that's ever really the case. Nevertheless, if he were speaking those words to us – to RB Builders – today, in our circumstances, I think what he would really be saying is that we cannot be afraid of confronting current reality.

"In his best-selling book *Good to Great*, author Jim Collins explains the Stockdale Paradox[13], and makes the point that great companies have the ability to confront the brutal facts of their current situation without ever losing faith in the ultimate outcome; the belief that that they will prevail – that they will win – in the end, regardless of the degree or length of the difficulties.

"Throughout these sessions – in fact, from the time I first started working with this company – I have encouraged you to start everything on the basis of an assessment of current reality. That's where it begins – with a realistic assessment of where you currently stand.

"See the situation exactly as it is, and for exactly what it is.

"Merom Klein wrote *The Courage to Act*, with Rod Napier.[14] In it, he talks about five factors required to find the courage to stay in business and build a successful enterprise: *Purpose* (the courage to pursue lofty goals); *Will* (the courage to inspire optimism and promise); *Rigor* (the courage of discipline); and, *Risk* (the courage to empower, trust, and invest in relationships).

---

[13] *Good To Great*, Jim Collins (Harper Collins, 2001). Collins recounts this duality (confront reality, unwavering faith in the ultimate outcome) known as the Stockdale Paradox, based on the experiences of Admiral James Stockdale, the highest-ranked US military officer held as a POW during the Vietnam War, imprisoned in Hanoi for eight years.

[14] *The Courage to Act*, Merom Klein and Rod Napier (Davies Black Publishing, 2003).

"Important attributes, all. But, Klein says that it all starts with the attribute of *Candor*, the courage to speak and hear the truth.

"Truth and reality go together. Let them always frame whatever you do."

| | 24 Rounds | | | |
|---|---|---|---|---|
| Standard/Smart | Projects | % | | |
| Revenue | 24 | 100% | | |
| Cost of Sales | -18 | -75% | | |
| Gross Income | 6 | 25% | | |
| Operating Expense | | | | |
| Net Income | | | | |

"That's the projection using 'standard' or 'smart' Resources. With 'strong' Resources, there is 25% more capacity – average capacity of five instead of four – but no higher cost. That's because Resources are now in Cost of Sales, which determines the capacity of the system, but not its cost; that is a big difference between the previous versions of the game and this new version."

| | 24 Rounds | | | |
|---|---|---|---|---|
| Strong | Projects | % | | |
| Revenue | 30 | 100% | | |
| Cost of Sales | -22 | -73% | | |
| Gross Income | 8 | 27% | | |
| Operating Expense | | | | |
| Net Income | | | | |

"In a 24-round game using 'strong' Resources, the expected Revenue would be 30 projects/jobs; Cost of Sales at that level would be 22 projects/jobs, rounded down; expected Gross Margin would be eight projects/jobs. Rounding skews the percentages a little bit; COS is 73% rather than 75%, Net Margin is 27% instead of 25%, and so forth, but I think it is important to deal with an Operating Statement with whole projects/jobs, and without fractions.

"So far, so good."

"As a result of removing capacity cost from external resources, we can look at games with shorter durations", she said, completing the portions of the tables that were considered to be 'above the line'.

| | 24 Rounds | | 12 Rounds | |
|---|---|---|---|---|
| Standard/Smart | Projects | % | Projects | % |
| Revenue | 24 | 100% | 12 | 100% |
| Cost of Sales | -18 | -75% | -9 | -75% |
| Gross Income | 6 | 25% | 3 | 25% |
| Operating Expense | | | | |
| Net Income | | | | |

| Strong | 24 Rounds Projects | % | 12 Rounds Projects | % |
|---|---|---|---|---|
| Revenue | 30 | 100% | 15 | 100% |
| Cost of Sales | -22 | -73% | -11 | -73% |
| Gross Income | 8 | 27% | 4 | 27% |
| Operating Expense | | | | |
| Net Income | | | | |

"A game with 12 rounds played with 'standard' or 'smart' Resources would have expected Revenue of 12 projects/jobs, Cost of Sales of nine projects/jobs, and expected Gross Margin of three projects/jobs; a 12-round game played using 'strong' Resources would have expected Revenue of 15 projects/jobs, Cost of Sales of 11 projects/jobs, and expected Gross Margin of four projects/jobs. And, playing the game with shorter duration wouldn't be bad; in fact, it would speed games up."

"I have been thinking about what you said earlier, about the benefit of restoring Cost of Sales to the Income Statement", said the CEO. "For purposes of a Pipeline game™, what we are now saying is that the financial perspective of Throughput is the fractional value of Revenue known as Contribution; i.e., Revenue, less Cost of Sales.

"Since financial information in a Pipeline game™ is expressed in terms of projects/jobs, Throughput – or Gross Margin or Contribution – is all of the projects/jobs completed during the specified time period of a game, less a stipulated percentage that equates to 75% of those projects/jobs, always expressed in terms of projects.

"Correct?"

"I see where you are headed", said the intrepid, results-based consultant, as she transferred the Operating Statement tables to some available space on the conference room wall and freed a new page on the flipchart.

"We need to define terms", she said, as she wrote on the flipchart.

Revenue = "Projects Completed"
Cost of Sales = "Projects Invested"
Operating Expense = "Projects Consumed"
Net Income = "Net Projects Completed"
Inventory = "Projects-in-Process"

"The definitions help", said the CEO, studying the Operating Statement tables. "Okay, now to the second issue. How we are treating Operating Expense?"

"Operating Expense is the financial investment in a system's production capacity", said the intrepid, results-based consultant. "It represents all of the money spent turning Inventory into Throughput. Operating Expense should be stipulated – 'imposed on the system' is a better phrasing – as an expression of the number of projects/jobs 'consumed' in the pursuit of projects/jobs 'completed'.

"In other words, having the capacity to build houses costs the system a portion of the Revenue it generates; a building company's capacity 'consumes' the Revenue from x-projects to fund its operations. Operating Expense is a budget imposed, based on the cost of the internal capacity required to manage work-in-process; it is only indirectly related to expected project completions.

"Operating Expense is a non-variable cost, so it spans the duration of the game", she said, completing the tables. "For all practical purposes, it is a fixed cost incurred by the system, regardless of the number of projects completed; it is incurred, even if it results in an operating loss, which is exactly how we look at overhead."

| | 24 Rounds | | 12 Rounds | |
|---|---|---|---|---|
| Standard/Smart | Projects | % | Projects | % |
| Revenue | 24 | 100% | 12 | 100% |
| Cost of Sales | -18 | -75% | -9 | -75% |
| Gross Income | 6 | 25% | 3 | 25% |
| Operating Expense | -4 | -17% | -2 | -17% |
| Net Income | 2 | 8% | 1 | 8% |

| | 24 Rounds | | 12 Rounds | |
|---|---|---|---|---|
| Strong | Projects | % | Projects | % |
| Revenue | 30 | 100% | 15 | 100% |
| Cost of Sales | -22 | -73% | -11 | -73% |
| Gross Income | 8 | 27% | 4 | 27% |
| Operating Expense | -4 | -14% | -2 | -14% |
| Net Income | 4 | 13% | 2 | 13% |

"Important to understand: Operating Expense is an amount – measured as the number of 'projects consumed' – that will support the necessary amount of work-in-process; it will support the number of 'projects-in-process' required by the system."

"I agree", said the CEO, still studying the tables. "Operating Expense imposed at four projects/jobs 'consumed' in a 24-round game, and two projects/jobs 'consumed' in a 12-round game? That's about right. I think Operating Expense imposed as a budget in that range – at 14% to 17% of Revenue – fits the profile of homebuilding."

"These changes raise another question", he said. "In switching from work performed by internal resources to work performed by external resources – and having those resources now define the capacity of the system, but no longer its cost – have we changed the visual image of homebuilding production?

"And, I'm thinking specifically about what determines the capacity of the pipeline and what determines the cost of the pipeline.

"The capacity of the pipeline has always been defined as the number of closings RB Builders could generate with a finite, planned, and controlled level of work-in-process", replied the intrepid, results-based consultant. That doesn't change. The cost of the pipeline has always been defined as a composition of 'Operating Expense and Resources'. I think the understanding remains unchanged, in principle, but the effect might be different, because we are now making a distinction between external resources and internal resources.

"Expanding that thinking, we are now making a distinction between direct variable costs attributable to external resources accounted for above-the-line in Cost of Sales, and indirect non-variable costs attributable to internal resources and overhead accounted for below-the-line as part of Operating Expense.

"If RB Builders was a manufacturing operation, or a vertically integrated homebuilding operation, the cost of its production pipeline would be more fixed, more concrete. That is not the case; RB Builders has an outsourced production system – a system where the overwhelming majority of the resources are external. In addition to the indirect non-variable costs of internal resources and overhead reflected in Operating Expense, RB Builders also has to consider the cost associated with establishing, maintaining, coordinating, scheduling, and supervising all of the external resources.

"It is not an insignificant requirement; it is a more difficult requirement to visualize, to quantify, to comprehend its significance.

"It adds to the character of Operating Expense more than it does Cost of Sales, because RB Builders is going to incur most of this cost as overhead; whether it is direct or indirect, it's mostly a non-variable cost.

"We are okay with that understanding as the cost of the pipeline."

"We have dealt with the Resources", said the intrepid, results-based consultant, transferring the sheet on terms to the conference room wall and freeing a new sheet on the flipchart. "We have dealt with the Operating Expense question. We have an Income/Operating Statement. Let's look at the formulas.

"How would you calculate Productivity?"

"We had been measuring Productivity as Throughput ÷ Operating Expense, but restoring Cost of Sales allows us to choose between Revenue and Gross Income as system output", said the CEO. "I think Productivity works better with the output measure tied to Revenue, not Gross Income; that's the traditional way to express the measure; honestly, it's the operating measure to which we have paid the least attention.

"Let's say Productivity = Revenue ÷ Expense, and we are dealing with projects/jobs as the unit of measure. So, in a 12-round game played with 'standard' or 'smart' Resources, it would be 12 ÷ 2 = 6 projects; if the game used all 'strong' Resources, it would be 15 ÷ 2 =

7.5 projects. It's an abstract number, but Productivity has always been an industry benchmark."

The intrepid, results-based consultant added the calculation to the flipchart.

Productivity = Revenue ÷ Operating Expense

Productivity = 12 ÷ 2
Productivity = 6 projects
        or
Productivity = 15 ÷ 2
Productivity = 7.5 projects

Productivity = Projects Completed ÷ Projects Consumed

"In Pipeline-speak, that makes Productivity = Projects Completed ÷ Projects Consumed", she said.

"What about Inventory Turn?"

"Again, the more useful, more acceptable association is going to lie with Revenue", said the CEO. "And – by the way – whenever we say 'Inventory', we are still talking about average projects/jobs-in-process for some specified period of time.

"So, it would be Inventory Turn = Revenue ÷ Inventory. In a 12-round game played with 'standard' or 'smart' Resources, it would be 12 ÷ 4 = 3; it should require work-in-process of four projects/jobs, meaning we should have a 3x turn; turn Inventory three times a year, three times during a game. In a game where all of the Resources were 'strong', it would be 15 ÷ 5 = 3; I think it would still be a 3x turn, because you would expect Resources with 25% more capacity to require 25% more projects/jobs-in-process.

"In Pipeline-ese, Inventory Turn would be Projects Completed ÷ Projects-in-Process.

"That's a lot more intuitive result than the previous game. Since they are reciprocals, I think that will be true of Cycle Time, as well."

The intrepid, results-based consultant added to the flipchart sheet.

Inventory Turn = Revenue ÷ Inventory
Inventory Turn = 12 ÷ 4; 15 ÷ 5
Inventory Turn = 3x
Inventory Turn = Projects Completed ÷ Projects-in-Process

"Speaking of which, how do we calculate the duration of the four phases in the process?", asked the CEO. "How do we calculate Cycle Time?"

"We use Little's Law, same as before. The formula for cycle time needs to be (Inventory ÷ Revenue) x $Days_n$", she said. "In a typical calendar-based cycle time calculation, the $Days_n$ part is still the number of days in the period we are calculating.

"In a Pipeline game™, recall we are dealing with number of rounds as the measure of duration, not the number of days. We take that outcome and relate it to rounds per calendar event, and then to the number of days we associate with that event.

"In a standard 12-round game, projected Revenue is 12 completed projects/jobs and projected Inventory is four projects/jobs. In order to calculate cycle time for the entire game – which would also be the equivalent of the calculation for the entire year – it would be 4 ÷ 12; take that quotient – 0.333 – and multiply it by 12; 0.333 x 12 = 4 rounds.

"The anticipated duration of a project/job would be four rounds, which is also how we would characterize the cycle time expected of the system. That's expected cycle time in a game played with 'standard' or 'smart' Resources; in a game played with 'strong' Resources, it has the same dynamic occurring as Inventory Turn, because Resources with higher capacity require more projects/jobs-in-process.

"In terms a 12-round Pipeline game™ uses, Cycle Time = (Projects-in-Process ÷ Projects Completed) x Rounds"

She added to the flipchart, as she spoke.

Cycle Time = (Inventory ÷ Revenue) x $Days_n$
Cycle Time = (4 ÷ 12) x 12; (5 ÷ 15) x 12
Cycle Time = 0.333 x 12
Cycle Time = 4 rounds

Cycle Time = (Projects-in-Process ÷ Projects Completed) x $Rounds_n$
Cycle Time = (Projects-in-Process ÷ Projects Completed) x 12

"And, as we know, in a 12-round game, each round represents a month, so four rounds represents four months", she said. "In a 24-round game, we would have to make the time adjustment. For ease-of-calculation, we use a 360 day year, which makes a month 30 days; a four-month duration is 120 days. And, that corresponds to the reciprocal of the 3x Inventory Turn we had calculated. 120 day durations and 3x turns look a lot more like a homebuilding operation."

"And, can we get away from consideration of a 24-round game?", asked the CEO. "We have the ability to run 24-round games if we choose, but I think we would cover more ground, learn lessons faster, if we only ran games with 12 rounds. I know a 12-round game magnifies variation, but we will learn more, faster."

"I agree", said the intrepid, results-based consultant. "If necessary, we can go back to a 24-round game. But, from this point forward, anything we talk about refers to a game played over 12 rounds.  We have covered the changes to operating performance measures.  What changes do we need to make to the business outcome measures?

"How should we calculate Net Income?  Net Margin?"

"Because we have restored Cost of Sales, the calculation of Net Income now looks a lot more like an Income Statement", replied the CEO.  "In the terms we apply to a Pipeline game™, Net Income = (Projects Completed - Projects Invested) - Projects Consumed.  Net Income is captured by a new Pipeline term:  'Net Projects Completed'.

"In a 12-round game – sorry, in any game played with 'standard' or 'smart' Resources – expected Net Income would be (12 - 9) - 2 = 1 project;  in a game played with 'strong' Resources, expected Net Income would be (15 - 11) - 2 = 2 projects.

"Net Income Margin would be Net Income ÷ Revenue;  in other words, it would be Net Projects Completed ÷ Projects Completed.  In a game played with 'standard' or 'smart' Resources, expected Net Income Margin would be 1 ÷ 12 = 8.3%.

"In a game played with 'strong' Resources, expected Net Income Margin would be 2 ÷ 15 = 13.3%.

"Net Income Margin ranging from 8.3% to 13.3%.  Very recognizable.  I like it."

The intrepid, results-based consultant noted everything the CEO said on the flipchart.

Net Income = Revenue - Cost of Sales) - Operating Expense

Net Income = (12 – 9)- 2
Net Income = 1 project
        or
Net Income = (15 – 11)- 2
Net Income = 2 projects

Net Projects Completed = (Projects Completed - Projects Invested) - Projects Consumed

Net Margin = Net Income ÷ Revenue

Net Margin = 1 ÷ 12
Net Margin = 8.3%
        or
Net Margin = 2 ÷ 15
Net Margin = 13.3%

Net Margin = Net Projects Completed ÷ Projects Completed

"What about Return on Invested Assets?", asked the intrepid, results-based consultant.

"As you told us: 'Learn this progression'", said the CEO, smiling and gesturing for the marker. "As now modified, of course."

The CEO removed the Net Margin sheet and posted it to the conference room wall, and started a new sheet on the flipchart.

ROIA = (Net Income ÷ Revenue) x (Revenue ÷ Inventory)
ROIA = Return on Sales x Inventory Turn
ROIA = Margin x Velocity

"In a game played with either 'standard' or 'smart Resources, we should be able to produce an 8.3% Net Margin and a 3x Inventory Turn", he explained, as he continued to write. "In a game played with all 'strong' Resources, we should get a 13.3% Net Margin and still generate a 3x Inventory Turn.

"Using the DuPont identity for ROIA, that combination of margin and velocity should produce a 25% to 40% Return on Invested Assets."

ROIA = (1 ÷ 12) x (12 ÷ 4)
ROIA = .083 x 3
ROIA = .249
ROIA = 24.9%, rounded to 25%
          or
ROIA = (2 ÷ 15) x (15 ÷ 5)
ROIA = .133 x 3
ROIA = .399
ROIA = 39.9%, rounded to 40%

"I just have to say it: a 25% to 40% Return on Invested Assets looks a lot more like the economic return that a homebuilding company would generate than the ROIA that the previous, manufacturing-oriented game produced. The only problem we are going to see will come from margins that are much narrower in the new game. Those more realistic margins are going to leave less room."

"Less room for what?", asked the intrepid, results-based consultant as she found another marker to complete the CEO's list.

ROIA = (Net Projects Completed ÷ Projects Completed) x (Projects Completed ÷ Projects-in-Process)

"I think the new game is going to offer fewer lessons", said the CEO. "Perhaps it's all the same lessons, but learned faster. I think we will learn as much as we did before, but learn it in fewer games. We just have to be careful with it.

"Here is one example. The system's production capacity is now determined by the work rate of external resources that are direct, variable costs: when they were internal Resources allocated/behaving as indirect, non-variable costs, hiring a 'strong' Resource had a cost factor. As you noted earlier: the fact is, we are not concerned with wasting the capacity of Resources that are direct, variable costs; retaining their loyalty, yes, but that doesn't affect the Operating Statement."

"And, consider this: we have started stipulating and imposing costs, both the direct, variable costs associated with Cost of Sales, and the indirect, non-variable costs associated with Operating Expense; these are now budgeted costs, their nature doesn't change; consequently, there are fewer decisions to make."

"I see your points", she said. "We will have to design the new simulations, with that understanding in mind. I know we will need a new scorecard that reflects these changes, and the new Operating Statement."

"Regarding the Operating Statement", said the CEO, pointing towards the tables and ticking through the line items. "I'm comfortable with all of the ratios, as being reflective of a homebuilding operation. Cost of Sales. Gross Margin. Operating Expense. Net Income Margin. It's good, but very unforgiving."

"With these changes, we are losing some of the pure connection – the consistency – we had with treating Throughput, Inventory, and Operating Expense as 'the only three things' that happen to money", said the intrepid, results-based consultant. "However, as long as we are consistent with the terms we use to connect operating performance to business outcomes, these changes will work, and the game will be easier to understand.

"One of the benefits of the change is that we have made Throughput synonymous with what it really is: Gross Income or Contribution. We won't have to fight the counter-intuitiveness of that relationship any longer."

She wrote it on the erasable board.

Throughput = Revenue - (Revenue x .75)
Throughput = Revenue - Cost of Sales
Throughput = Projects Completed - Projects Invested

"We don't need to install breakeven as a required business measure, but the game allows us to consider it", said the CEO. "Breakeven clarifies the relationship between Cost of Sales and Operating Expense. In a game with either 'standard' or 'smart' Resources, Operating Expense 'consumes' two projects per game, and Cost of Sales diminishes each completed project by 75%; breakeven is achieved at eight completed projects."

"Since everything is measured in terms of 'projects', one of the stipulations has to be no fractional measures", said the intrepid, results-based consultant. "It needs to be just like the Operating Statement tables; scoring has to be in 'whole projects', which will produce minor discrepancies in percentages, but we can live with it."

"Are you ready?", asked the CEO. "Can I bring everyone back?"

"It's only 1:00", she said. "I have everything I need, but give me the remaining 30 minutes to get all of it pulled together."


When the team was reassembled, the intrepid, results-based consultant explained the game, with the changes.

"I am going to explain the changes we have made to the game", she said. "When you were learning the game, it was important to reveal the different elements gradually, so that you could move from where you were, productionally-speaking, to where you needed to be."

One superintendent turned to another superintendent, and asked, "Really? 'Productionally'? Is that even a word?"

The other superintendent shrugged and replied, "It is now."

"Of course, it's a word", said one of the sales representatives. "Productionally-Transmitted Diseases? PTD? The scourge of all superintendents?"

"You have all run the production simulations numerous times", said the intrepid, results-based consultant. "You know the principles well, but these changes will give you a more realistic perspective for how the principles apply in a production environment specific to homebuilding.

"Overall, these changes represent an improvement to the Pipeline game™.

"More than anything else, this is the future, the version we will teach to others, to new teammates."

She pulled up the first, and then, a second PowerPoint slide.

"So – here is the information on a balanced capacity system and on a system with unbalanced capacity. Both systems should look very familiar, but note that we have shortened the duration of the game from 24 rounds to 12 rounds."

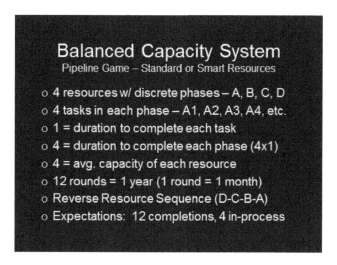

"There are some new terms you have to learn and use, and we have cemented some of the terms we used before. The new terms were required when we restored Cost of Sales to the Operating Statement, also necessary when we separated Resource capacity from Resource cost."

She pulled up the next slide.

**Terms and Definitions**
(expressed in projects)

o  Revenue = "Projects Completed"
o  Cost of Sales = "Projects Invested"
o  Operating Expense = "Projects Consumed"
o  Net Income = "Net Projects Completed"
o  Inventory = "Projects-in-Process"

"This was our previous method of measuring operating performance and business outcomes", she said, pulling up one of the earlier slides. "Everything was oriented towards the question: 'what happens to money?'."

**The Impact on Profitability and Economic Return**

Productivity = $Throughput \div Expense$
Inventory Turn = $Throughput \div Inventory$
Net Income = $Throughput - Expense$
ROA = $(Throughput - Expense) \div Inventory$
ROA = $(Net Income \div Throughput) \times (Throughput \div Inventory)$
Cycle Time = $(Inventory \div Throughput) \times days$

"Still a valid and important question, but now, the operational and financial scoring terms in a game are always related to projects – Projects Completed, Projects Invested, Projects Consumed, Projects-in-Process.

"The use of those terms allow us to measure the following operating measures . . . ", she said, pulling up the next slide.

## Operating Measures
### (expressed in projects)

o Productivity = Projects Completed ÷ Projects Consumed

o Inventory Turn = Projects Completed ÷ Projects-in-Process

o Cycle Time = (Projects-in-Process ÷ Projects Completed) x 12

" . . . and the following measures of business performance", she continued, pulling up another slide. "As you see, we cover all of the main line-item categories on the Operating Statement, based on a Contribution Income Statement, and using a variable costing approach to cost allocation."

## Financial Measures
### (expressed in projects)

o Revenue = Projects Completed

o Gross Income (Throughput) = Projects Completed - Projects Invested

o Net Income = (Projects Completed - Projects Invested) - Projects Consumed

o Net Income Margin = Net Projects Completed ÷ Projects Completed

o ROA = Net Projects Completed ÷ Projects-in-Process

She pulled up another slide. "As you know, we look at economic return as Return on Invested Assets, and we look at ROIA as a co-equal function of margin x velocity, using the expanded view provided by the DuPont identity. And, as you know, the DuPont identity looks at ROA as Return on Sales x Asset Turn."

## Financial Measures
### (2)

o DuPont ROA = (Net Projects Completed ÷ Projects Completed) x (Projects Completed ÷ Projects-in-Process)

o DuPont ROA = (Net Projects Completed ÷ ~~Projects Completed~~) x (~~Projects Completed~~ ÷ Projects-in-Process)

o DuPont ROA = Return on Sales x Asset Turn

o DuPont ROA = Margin x Velocity

"Resources", she said categorically, transitioning through the next slide. "No changes to the Resources themselves, except Resources no longer determine both the capacity and the cost of the system; they determine capacity only; in that regard, 'standard', 'smart', and 'strong' Resources still determine capacity the same as before; in terms of variation, they have the same probabilities, the same distributions."

## Resources

- Standard Resource: sufficient, capable; average capacity = 4 (1-2-3-4-5-6); P/D: 60%/20%

- Smart Resource: sufficient, capable, less variation; average capacity = 4 (1-2 = 3, 3-4 = 4, 5-6 = 5); P/D: 67%/33%

- Premium Resource: more than sufficient, less variation; average capacity = 5 (3-4 = 4, 2-5 = 5, 1-6 = 6); 67%/33%

The intrepid, results-based consultant pulled up her final slide.

**Operating Statement**
unbalanced capacity, Standard, Smart, Premium
resources, in projects

| | Std./Smt. | % | Prem. | % |
|---|---|---|---|---|
| Revenue | 12 | 100% | 15 | 100% |
| COS | -9 | -75% | -11 | -73% |
| GI | 3 | 25% | 4 | 27% |
| OE | -2 | -17% | -2 | -14% |
| NI | 1 | 8% | 2 | 13% |

"Finally, here is the new Operating Statement, representing expectations, and, therefore, a factor in your budgets. It references the unbalanced capacity system, but it applies to both balanced and unbalanced capacity systems; this Operating Statement considers all of the Resources you might have available to you in various circumstances.

"Cost of Sales is 'stipulated' for the system, Operating Expense is 'imposed' on the system. Think of both as being budgeted: Cost of Sales as a function of job budgets, calculated at 75% of Revenue or Projects Completed; and, Operating Expense as a function of budgeted overhead that will 'consume' two projects per game.

"So, we have Projects Invested, which is Cost of Sales, and we have Projects Consumed, which is Operating Expense.

"Let's see how all of these changes impact how you manage homebuilding production, how you manage a homebuilding production system", she said, passing around a new set of score sheets. "We are going to reduce the number of production situations we simulate, in order to make the outcomes more stark; the comparative results scorecards have been updated.

"There are new game score sheets for you to use, which provide the new formulas, and the imposed Cost of Sales and Operating Expense numbers. Game 1; actually, let's call it Game 1B, to differentiate it from earlier versions. We have a balanced capacity system, like this."

She pulled up the balanced capacity system slide on her notebook.

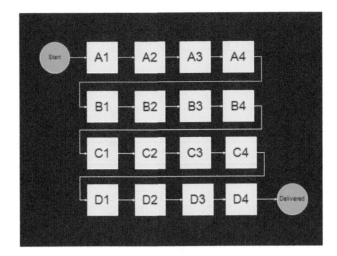

"We are back to Square 1: a production system with balanced capacity; four discrete processes, four tasks in each process; four 'standard' Resources to perform the work in each process; a 2-3-4-5-6 dice roll profile, average roll of four, therefore an average Resource capacity of four.

"So – what are the projections? Start with Revenue, with what we term 'Projects Completed'. How many?"

"We are using 'standard' Resources, which means we are not showing any partnering improvements", offered one of the superintendents. "We have a system with balanced capacity, which means we aren't showing any process improvements. Four processes. Four tasks in each process. An average capacity of four across all Resources. We should be able to complete one project/job per round; 12-round game, we should have Projects Completed of 12 projects."

The intrepid, results-based consultant started a list on the erasable board.

Revenue (Projects Completed) = 12 projects
Cost of Sales (Projects Invested) = 12 projects x .75 = 9 projects
Gross Income (Contribution Margin, Throughput) = 12 – 9 = 3 projects
Operating Expense (Projects Consumed) = 2 projects; 'fixed cost'
Net Income (Net Projects Completed) = (12 – 9) – 2 = 1 project
Net Margin = 1 ÷ 12 = 8.3%

"Net Margin gives us the Return on Sales component of economic return, but we need some operating projections to calculate Asset Turn component. Give me Productivity, Inventory Turn, and Cycle Time."

"Productivity is simple", replied the CFO. "It's 'Projects Completed' divided by 'Projects Consumed'; that's a Productivity rate of six projects; so, a productivity rate of 6.0. But, I need to know the average number of projects/jobs-in-process we will be carrying, before I can calculate either Inventory Turn or Cycle Time."

"If the game has a balanced capacity system and it involves either 'standard' or 'smart' Resources, we need to carry four 'Projects-in-Process' throughout a game", answered the VP of Construction. "Inventory Turn is 'Projects Completed' divided by 'Projects-in-Process', which should produce an Inventory Turn of 3x, because we should have 12 completed projects/jobs and four jobs-in-process.

"Cycle Time is the reciprocal of Inventory Turn; 'Projects-in-Process' divided by 'Projects Completed', adjusted for the number of rounds; so, four 'Projects-in-Process' divided by 12 'Projects Completed', times 12 rounds.

"Cycle Time should calculate to four rounds; when you think about it, each project/job takes four rounds to move through the A, B, C, and D processes; since a round is equivalent to 30 days, Cycle Time should be 120 days."

The intrepid, results-based consultant continued the list.

Productivity (Projects Completed ÷ Projects Consumed) = 12 ÷ 2 = 6 projects
Inventory Turn (Projects Completed ÷ Projects-in-Process) = 12 ÷ 4 = 3x

Cycle Time = (Projects-in-Process ÷ Projects Completed) x 12;
Cycle Time = (4 ÷ 12) x 12 = 4 periods;
Cycle Time = 4 periods x 30 days = 120 days

"Now that we have both Return on Sales and Asset Turn, we can project Return on Invested Assets", said the CFO. "Net Income Margin – which is Return on Sales – is expected to be 8.3%; Asset Turn – which is calculated from Inventory Turn – is expected to be 3x. Since we presume all of our assets are invested in projects/jobs-in-process, we would expect to generate an ROIA of 24.9%.

"25%, rounded."

The intrepid, results-based consultant completed the list.

ROIA = (Projects Completed – Projects Invested) – Projects Consumed x (Projects Completed ÷ Projects-in-Process);
ROIA = Return on Sales x Asset Turn;
ROIA = .083 x 3.0 = 24.9%; rounded to 25%
ROIA = Margin x Velocity

"Congratulations", she said, smiling. "You've been away from this stuff for more than an hour and a-half, and you still remember it." Then she spoke more seriously. "You *have* learned a tremendous amount about managing production during our many previous sessions. But, now I need you to lay that knowledge aside, and to run this new progression of production scenarios, as if you had never seen any of it.

"Play the game like you are still in your silos;  play it like you are being bonused on individual job performance.

"Game 1B.  Play it."

## GAME 1B

Game 1B was played with all 'standard' Resources, using a system where the workload was balanced across all of the processes;  the four Resources were all superintendents;  Necessary WIP ('Projects-in-Process') was an average of four projects/jobs-in-process, but there was no mandated Minimum WIP or Maximum WIP.

"How did you do?", asked the intrepid, results-based consultant, pointing towards the large comparative results scorecard she had drawn.

"Not very well", replied the CFO.  "Before we go there, we need to revisit the rule about the 'rounding to whole projects' restriction in the scoring.  Zero is not reality;  we can report in whole projects, but, technically, we need to allow fractional scoring;  a 12-round game is very short;  whole project outcomes actually become a distortion of reality;  fractions make a difference, in terms of what they help us see.

"I know that a partially complete or fractional project/job is not a marketable outcome, not a marketable reality.  But, neither are defined periods of time a reality;  time is continuous;  a year is only a year because we choose to say that it is.  There is partially finished work entering most periods, and partially complete work at the end of most periods, because that's what work-in-process is.

"So – I suggest we allow the discussion of outcomes of simulations, in terms of both whole and fractional projects, particularly when playing a 12-round game."

The intrepid, results-based consultant glanced around the conference room, as she mulled the request.  Without elaborating on the request or commenting further, she simply said, "Okay."

## GAME 1B

| | G-1B | | | | | | | |
|---|---|---|---|---|---|---|---|---|
| Description<br>Std. = 4<br>Smart = 4<br>Strong = 5 | Balanced,<br>4 Std.,<br>Necessary<br>WIP = 4 | | | | | | | |
| | | | | | | | | |
| **Budgeted** | | | | | | | | |
| Revenue | 12 | | | | | | | |
| P-I-P | 4 | | | | | | | |
| OE | 2 | | | | | | | |
| P-Rate | 6.0 | | | | | | | |
| Inv. Turn | 3X | | | | | | | |
| CT (days) | 120 | | | | | | | |
| NI | 1 | | | | | | | |
| NI % | 8.3% | | | | | | | |
| ROIA | 25% | | | | | | | |
| | | | | | | | | |
| Actual | | | | | | | | |
| Revenue | 9 | | | | | | | |
| P-I-P | 4 | | | | | | | |
| OE | 2 | | | | | | | |
| P-Rate | 4.5 | | | | | | | |
| Inv. Turn | 2.3X | | | | | | | |
| CT (days) | 160 | | | | | | | |
| NI | .25 | | | | | | | |
| NI % | 2.8% | | | | | | | |
| ROIA | 6.3% | | | | | | | |
| | | | | | | | | |
| Resource A | 4.3 | | | | | | | |
| Resource B | 3.7 | | | | | | | |
| Resource C | 4.2 | | | | | | | |
| Resource D | 3.6 | | | | | | | |
| | | | | | | | | |

"In Game 1B, we completed nine projects", reported the CFO. "Nine, not 12. If someone said they met their expected output three-quarters of the time, it wouldn't sound that terrible; but, there was not a single period in which we exceeded expectations, to offset the periods when we didn't meet expectations.

"The one-quarter of the rounds where we didn't meet expected output killed our performance for the entire game. At 75% of Projects Completed, Cost of Sales (expressed as Projects Invested) is a variable cost, but it reduced Throughput from an anticipated three projects to

2.25 projects – only two projects, if we round to whole projects – dooming us to barely absorb the non-variable cost that comprises our Operating Expense; 'Projects Consumed' is a good term for it."

"Net Income, or Net Projects Completed, was .25 projects – really, zero Net Projects Completed, if we rounded to the nearest whole project. Therefore, Net Income Margin was 2.25%, not the anticipated 8.3%.

"We controlled work-in-process fairly well, averaging 4.1 Projects-in-Process, versus the expected 4.0 Projects-in-Process; during the game's 12 rounds, Projects-in-Process ranged between three and five projects.

"In terms of operating performance, Productivity was 4.5 projects, not the projected six projects. Inventory Turn was 2.25x, rather than 3x; Cycle Time was 5.3 periods, not the anticipated four periods, meaning Cycle Time was 160 days, not 120 days.

"As a result, ROIA was only 6.3%, instead of the expected 25%.

"The four Resources averaged work of 3.95, right at the expected four projects per period; two averaged above four; two averaged below four. Periods 3, 4, and 9 were miserable circumstances, in which all of the Resources combined for work rates that were significantly below average; that was the quarter of work periods in which we missed completed projects.

"That was the difference in the game. We could never make up the deficit; capacity was too balanced, and there was no reserve capacity."


## GAME 2B


"Let's reduce the variation we experience with 'standard' Resources", said the intrepid, results-based consultant. "In real life, reducing variation and uncertainty is difficult, persistent work. It doesn't necessarily create more capacity or make Resources more productive, but it does make them more consistent, more predictable.

"In the previous version of the Pipeline game™, we introduced 'smart' Resources, which have same capacity as 'standard' Resources – average roll of four – but they do it with a less variable profile, a 3-4-5 profile, versus a 2-3-4-5-6 profile.

"Let's run Game 2B using all 'smart' Resources; the system has balanced capacity across all of its Resources, and a balanced workload in all of its processes; it's a balanced capacity system."

Game 2B was played with all 'smart' Resources in a system that had balanced capacity; the Resources were again all superintendents; Necessary WIP ('Projects-in-Process') – the required amount work-in-process – was still regarded as an average of four projects/jobs-in-process, with no mandated Minimum WIP or Maximum WIP.

## GAME 2B

| Description<br>Std. = 4<br>Smart = 4<br>Strong = 5 | G-1B<br>Balanced, 4 Std., Necessary WIP = 4 | G2B<br>Balanced, 4 Std., Necessary WIP = 4 | | | | | | | |
|---|---|---|---|---|---|---|---|---|---|
| | | | | | | | | | |
| **Budgeted** | | | | | | | | | |
| Revenue | 12 | 12 | | | | | | | |
| P-I-P | 4 | 4 | | | | | | | |
| OE | 2 | 2 | | | | | | | |
| P-Rate | 6.0 | 6.0 | | | | | | | |
| Inv. Turn | 3x | 3x | | | | | | | |
| CT (days) | 120 | 120 | | | | | | | |
| NI | 1 | 1 | | | | | | | |
| NI % | 8.3% | 8.3% | | | | | | | |
| ROIA | 25% | 25% | | | | | | | |
| | | | | | | | | | |
| **Actual** | | | | | | | | | |
| Revenue | 9 | 10 | | | | | | | |
| P-I-P | 4 | 5.5 | | | | | | | |
| OE | 2 | 2 | | | | | | | |
| P-Rate | 4.5 | 5.0 | | | | | | | |
| Inv. Turn | 2.3x | 1.8x | | | | | | | |
| CT (days) | 160 | 198 | | | | | | | |
| NI | .25 | .50 | | | | | | | |
| NI % | 2.8% | 5.0% | | | | | | | |
| ROIA | 6.3% | 9.0% | | | | | | | |
| | | | | | | | | | |
| Resource A | 3.6 | 4.1 | | | | | | | |
| Resource B | 4.2 | 3.9 | | | | | | | |
| Resource C | 3.7 | 4.1 | | | | | | | |
| Resource D | 4.3 | 3.9 | | | | | | | |
| | | | | | | | | | |

"Results?", asked the intrepid, results-based consultant, moving back to the comparative results scorecard.

"Overall, Game 2B was better than Game 1B", said the CFO.

"We had 10 Projects Completed, so we had Revenue of 10 projects. Cost of Sales is a variable cost at 75% of Revenue, so Projects Invested was 7.5 projects. Throughput – i.e., Gross Income – was therefore 2.5 projects.

"Operating Expense is an imposed budget of two Projects Consumed per game – in other words, a non-variable cost – which made Net Projects Completed, i.e., Net Income, equal to .5 projects. Net Income Margin was 5.0%, which is calculated as .5 Net Projects Completed divided by 10 Projects Completed.

"We were not as good at controlling work-in-process in Game 2B, averaging 5.5 Projects-in-Process versus the 4.0 Projects-in-Process that is considered adequate; Projects-in-Process ranged between three and six projects; eight of the 12 rounds had six Projects-in-Process, three carried five Projects-in-Process; the only round ending with less than four Projects-in-Process was Round 1.

"In terms of operating performance, Productivity improved to 5.0 projects, still not the projected six projects. As a result of the higher average Projects-in-Process, Inventory Turn decreased to 1.8x, moving away from the expected 3x; Cycle Time was therefore worse, at 6.6 periods, instead of the anticipated four periods, equating to 198 days instead of 120 days.

"ROIA was 9.0%, purely on the higher Return on Sales component; Asset Turn was lower. So – we had improved economic return, on better margin, despite worse velocity; Return on Invested Assets in Game 2B was higher than in Game 1B, but neither outcome achieved the expected level of 25%.

"The average work rate across the four Resources was exactly 4.0 per period; again, two Resources were slightly above, two were slightly below. In this particular case, variation was considerably less with 'smart' Resources, so this game was a good picture of the advantage of reducing variation."

"Statistically, variation in Game 1B was more than three times that in Game 2B", added the VP of Construction. "I have to say, though, we have previously seen greater instances of variation than in either of these two games. In a real production situation, this variation would be interesting forensics for applying PDCA, or kaizen.

"Moreover, in nine of the 24 per periods in which Resource C and D worked, the amount of work they could have performed was below average; that's 38%. Plus, there was a series of six periods in the middle of the game, in which Resource D worked at a below-average rate 83% of the time."

"The difference between making/not making budget came down to two of the forty-eight rolls in the game, less than five percent of the work effort", said the CFO. "Those two rolls resulted in the two periods in which we failed to generate a completed project. Once again, we never had the capacity to make up a deficit."

GAME 3B

"Okay, let's take another look at what happens whenever we add production capacity to a balanced system", said the intrepid, results-based consultant.

"From previous games, you are familiar with 'strong' Resources. These are Resources that have the capacity to perform more work than what the system requires; 'strong' Resources have more capacity than 'standard' or 'smart' Resources, and they have the same profile of reduced variation as 'smart' Resources; they are both 'smart' and 'strong'."

She rummaged through the flipchart sheets from previous games until she found the sheet she wanted, making one small change. "As you recall . . . "

Standard: 4 = avg. work completed/round (1-2-3-4-5-6)
Smart: 4 = avg. work completed (1 or 2 = 3; 3 or 4 = 4; 5 or 6 = 5)
Strong: 5 = avg. work completed (3 or 4 = 4; 2 or 5 = 5; 1 or 6 = 6)

"Like 'smart' Resources, we let 'strong' Resources roll anything – roll everything – on the dice face, 1-2-3-4-5-6", she said. "A roll of 3 or 4 = 4; a roll of 4 or 5 = 5; a roll of 1 or 6 = 6, logical, easy to remember. That gives 'strong' Resources an average roll/workrate/capacity of 5, versus 'standard' and 'smart' Resources, which have average roll/workrate/capacities of 4.

"Moreover, 'strong' Resources have the same 33%-33%-33% probability distribution as 'smart' Resources, which is an improvement over the 20%-20%-20%-20%-20% distribution of a 'standard' Resource. So – while the 'standard' Resource has a 60% probability of rolling a 4 or higher, and the 'smart' Resource has a 67% probability of rolling a 4 or higher – the 'strong' Resource has a 67% probability for producing a workrate of 5 or higher.

"4-5-6, instead of either 3-4-5, or 2-3-4-5-6.

"Before we talk about the changes that 'strong' Resources impart to the Operating Statement, we need to consider how a decision to add – or to increase – Resource capacity changes the nature of the game.

"In the previous version of the game, Resources defined both the *capacity* and the *cost* of the system. In order to have the production simulations more replicate homebuilding production – as opposed to manufacturing – we restored Cost of Sales as a line item to the Operating Statement, and made Throughput synonymous with Gross Income, instead of Revenue.

"To be clear, every enterprise has some level of direct, variable cost, albeit in proportions that make for a variety of Operating Statements; Cost of Sales was missing altogether in the previous version of the Pipeline game™, and that made it look more like a manufacturing operation.

"When we restored Cost of Sales, it meant that Resources would continue to define the capacity of the system, but would no longer define its cost. Resources became external, meaning that the cost of those Resources became part of Cost of Sales, thus, became direct, variable costs.

"The changes we have made to costs – to both direct, variable costs and to indirect, non-variable costs – alters the decision-making on Resource capacity. Previously, if you increased the capacity of a Resource, you also increased its cost; it was worth more, therefore it cost you more; you paid for it, whether you used it or not.

"That is no longer the case. Resource cost is now a direct, variable cost; you pay for it because you used it, not because you had it.

"As we discussed, Cost of Sales is stipulated as a variable cost at 75% of each project's value, a very common ratio in homebuilding. Operating Expense, which represents all of the indirect, non-variable costs incurred to manage production – largely to supervise work-in-process – is imposed as a budget; Operating Expense 'consumes' two completed projects each game, regardless of how many Projects Completed, regardless of how many Projects-in-Process you carry.

"You are going to find this arrangement and relationship more self-regulating."

"That must be nice", said the VP of Sales. "Snap your fingers, and – poof – you have more capacity, but don't have to pay for it unless you use it. Variation: snap your fingers, and – poof – you have Resources that are instantly more predictable, more dependable."

"I think we are going to stipulate that the Resource A position is always filled by a sales representative", said the CEO. "Just keep making it rain, Chief, and leave the production-thinking to the F-150 and Silverado crowd."

"That is sometimes the problem with simulations", said the intrepid, results-based consultant. "It looks too easy. In reality, for RB Builders, squeezing more capacity out of external Resources – by definition, Resources that RB Builders does not control – is a profile in what we term Epic Partnering®; the same with reducing variation, with making Resources more predictable and dependable.

"Fostering epic relationships of mutual interest with your building partners and supply partners is hard, painstaking work. As we said, the RB-IPS provides both the process and the program for progressively transforming suppliers and subcontractors into true partners, into trusted allies, joined together by their shared, mutual interests.

"It's anything but easy.

"Now – let's discuss the changes that the use of 'strong' Resources imposes on the simple Operating Statement we use in these simulations", she said, pulling up one of the previous PowerPoint slides.

"Ignore the note advising you whether the system has balanced or unbalanced capacity. It refers to the relationship between workload capacity. The way we have designed these production simulations, the production system can be unbalanced by workload or by resource; the constraint inherent in an unbalanced system can be created by the process or a resource; that difference between workload or capacity is what determines whether the system is in a balanced state, or an unbalanced state.

"In this case, the type and mix of Resources determines capacity.

"In a 12-round game, the use of 'standard' or 'smart' Resources carries the expectation of 12 Projects Completed; using 'strong' Resources increases the expectation to 15 Projects Completed", she explained. "Regardless of the resource configuration, COS is stipulated – as a variable cost – at 75% of every project's value.

"Expressing the Operating Statement in terms of 'whole' projects requires us to make minor adjustments to percentages; odd numbers are a bit more difficult to work with than even numbers."

## Operating Statement
### unbalanced capacity, Standard, Smart, Premium resources, in projects

|         | Std./Smt. | %    | Prem. | %    |
|---------|-----------|------|-------|------|
| Revenue | 12        | 100% | 15    | 100% |
| COS     | -9        | -75% | -11   | -73% |
| GI      | 3         | 25%  | 4     | 27%  |
| OE      | -2        | -17% | -2    | -14% |
| NI      | 1         | 8%   | 2     | 13%  |

"Before you run Game 3B, give me your budget/projections for the operating performance measures and business outcomes, based on all 'strong' Resources", she said, selecting a marker, moving to an unused portion of the erasable board, and recording the team's projections.

"We should be able to complete 15 projects in this game", said one of the sales representatives, reading from the slide. "Gross Income should be four projects, and Net Income should be two projects.

"That's the extent of our contribution. Don't expect any more from us", she said. "The superintendents need to step it up."

'strong' = 5; 'smart' = 4; 5 ÷ 4 = 1.25
Revenue (Projects Completed) = 12 projects x 1.25 = 15 projects
Cost of Sales (Projects Invested) = 15 projects x .75 = 11 projects
Gross Income (Contribution Margin, Throughput) = 15 – 11 = 4 projects
Operating Expense (Projects Consumed) = 2 projects; 'fixed cost'
Net Income (Net Projects Completed) = (15 – 11) – 2 = 2 projects
Net Margin = 2 ÷ 15 = 13.3%

"Let's see;  more Projects Completed with the same number of Projects Consumed", said a superintendent, smiling.  "15 Projects Completed divided by two Projects Consumed; Productivity should increase to 7.5 projects."

$$Productivity = (Projects\ Completed \div Projects\ Consumed) = 15 \div 2 = 7.5\ projects$$

"I would expect the Projects-in-Process requirement to move in relationship to increases in capacity;  we have average capacity of five instead of four;  since we have 25% more capacity, we probably need 25% more to work on;  so, let's say average Projects-in-Process of five.  In that case, 15 Projects Completed divided by five Projects-in-Process would leave Inventory Turn unchanged at 3x."

$$Inventory\ Turn\ (Projects\ Completed \div Projects\text{-}in\text{-}Process) = 15 \div 5 = 3x$$

"Cycle Time is the reciprocal of Inventory Turn", he continued.  "So, five Projects-in-Process divided by 15 Projects Completed, times the number of periods.  That would be four periods, times 30 days per period.  Since they are reciprocals, it stands to reason that Cycle Time would also be unchanged:  four periods, 120 days."

$$Cycle\ Time = (Projects\text{-}in\text{-}Process \div Projects\ Completed) \times 12;$$
$$Cycle\ Time = (5 \div 15) \times 12 = 4\ periods;$$
$$Cycle\ Time = 4\ periods \times 30\ days = 120\ days$$

"In terms of Return on Invested Assets, Net Income Margin is the same as the Return on Sales, the margin component;  two Net Projects Completed divided by 15 Projects Completed;  Net Income Margin is 13.3%.  In the Pipeline game™, Inventory Turn is the equivalent of Asset Turn, the velocity component;  15 Projects Completed divided by five Projects-in-Process;  Asset Turnover is 3x.  So – ROIA should be 13.3% times three turns.  ROIA should be 40%, rounded."

$$ROIA = (Projects\ Completed - Projects\ Invested) - Projects\ Consumed \times (Projects\ Completed \div Projects\text{-}in\text{-}Process);$$
$$ROIA = Return\ on\ Sales \times Asset\ Turn;$$
$$ROIA = .133 \times 3.0 = 39.9\%;\ rounded\ to\ 40\%$$
$$ROIA = Margin \times Velocity$$

The intrepid, results-based consultant set the marker down and said, "Let's play Game 3B with four 'strong' Resources in a system with balanced capacity."

Game 3B was played with all 'strong' Resources in a system with balanced capacity; Resource A was filled by a sales representative, B, C, and D by superintendents;  Necessary WIP ('Projects-in-Process') was considered to be an average of five projects/jobs-in-process; however, there was no Minimum WIP or Maximum WIP requirement imposed, or any guidance provided.

"How did you do?", asked the intrepid, results-based consultant.

"Having the more robust Resources made a difference", said the CFO.  "At least, on the margin side.  We completed 14 projects/jobs, which enabled us to better leverage our Operating Expense.  Net Income – that is, our Net Projects Completed – tripled from the previous game;  since we also had higher Revenue – more Projects Completed – our Net Margin merely doubled.

"Since it is tied so tightly to the Operating Statement, as opposed to a Balance Sheet, our Productivity was good, 7.0, but not quite the projected 7.5.

"The velocity side was a different story.  Game 3B produced our worst Inventory Turn and Cycle Time performance, so far.  Inventory Turn was only 2.2x, and Cycle Time increased to 201 days;  that's a far cry from the 3x and 120 day performance we wanted."

"Why did it happen?", asked the intrepid, results-based consultant.

"We carried a lot more Projects-in-Process than we thought we would", replied the VP of Construction.  "We carried an average of almost eight projects/jobs, instead of the projected five, fifty percent more than we thought.  It seemed like 40% to 50% of the Projects-in-Process were always in the A Process, a long way from completion, and where it is harder to say whether it was doing us any good."

"What did you expect me to do?", asked the sales representative who had been Resource A.  "I start projects/jobs whenever I can.  That's my job;  and – by the way – that's how I get bonused.  There were times when I could have started more projects/jobs than I did;  I held off in some rounds."

## GAME 3B

| Description<br>Std. = 4<br>Smart = 4<br>Strong = 5 | G-1B<br>Balanced,<br>4 Std.,<br>Necessary<br>WIP = 4 | G2B<br>Balanced,<br>4 Smart,<br>Necessary<br>WIP = 4 | G3B<br>Balanced,<br>4 Strong<br>Necessary<br>WIP = 5 | | | | | |
|---|---|---|---|---|---|---|---|---|
| **Budgeted** | | | | | | | | |
| Revenue | 12 | 12 | 15 | | | | | |
| P-I-P | 4 | 4 | 5 | | | | | |
| OE | 2 | 2 | 2 | | | | | |
| P-Rate | 6 | 6 | 7.5 | | | | | |
| Inv. Turn | 3X | 3X | 3X | | | | | |
| CT (days) | 120 | 120 | 120 | | | | | |
| NI | 1 | 1 | 2 | | | | | |
| NI % | 8.3% | 8.3% | 13.3% | | | | | |
| ROIA | 25% | 25% | 40% | | | | | |
| | | | | | | | | |
| **Actual** | | | | | | | | |
| Revenue | 9 | 10 | 14 | | | | | |
| P-I-P | 4 | 5.5 | 7.8 | | | | | |
| OE | 2 | 2 | 2 | | | | | |
| P-Rate | 4.5 | 5.0 | 7.0 | | | | | |
| Inv. Turn | 2.3X | 1.8X | 2.2X | | | | | |
| CT (days) | 160 | 198 | 201 | | | | | |
| NI | .25 | .50 | 1.50 | | | | | |
| NI % | 2.8% | 5.0% | 10.7% | | | | | |
| ROIA | 6.3% | 9.0% | 23.6% | | | | | |
| | | | | | | | | |
| Resource A | 3.6 | 4.1 | 4.9 | | | | | |
| Resource B | 4.2 | 3.9 | 4.7 | | | | | |
| Resource C | 3.7 | 4.1 | 5.1 | | | | | |
| Resource D | 4.3 | 3.9 | 5.0 | | | | | |
| | | | | | | | | |

"You are Resource A, where projects/jobs enter the system", said the intrepid, results-based consultant. "How did you understand your role as pacemaker? How did you view the pace of starts you were expected to maintain?"

"You mean, other than being her greedy little self?", asked a superintendent.

"I felt like it was my responsibility, part of my job", said the sales representative, "to push as many projects/jobs as I could into the system; also to make sure there was always a project/job in the A4 (complete) task that Resource B could pull into his process; and to keep everyone busy, with projects/jobs in every task – in A1, A2, A3, and A4."

"Interesting use of production terms, 'push' and 'pull'", said the intrepid, results-based consultant. "You were pushing projects/jobs into the A process, while B was supposed to pull your completed projects/jobs into his process.

"Care to comment on the distinction?"

"Don't mind if I do", she replied. "First of all, there was never any stipulation about whether this production simulation was a push system or a pull system.

"Absent a mechanism that enforces pull, like rate of completions, or the load in front of a Constraint Capacity Resource relative to its capacity, Resources are going to push projects/jobs into the system, and they will do it according to a mechanism that determines more than just the order of starts.

"It will specify the rate of starts, as well."

"We will want to establish pull, as we go forward", said the intrepid, results-based consultant. "What else was worthy of note in Game 3B?"

"It is a fact that we carried more Projects-in-Process than we intended, and it showed in slower turns and longer duration", said the CFO. "We had a 10.7% Net Margin and a 2.2x Inventory Turn; using those measures as the Return on Sales and Asset Turn components of Return on Assets, we had an ROIA of 23.6%.

"That is the best economic return we have generated, so far, under this new version of the game, but still short of the estimated 40% ROIA.

"The periodic work rate across all four Resources averaged 4.9, slightly below the 5.0 we expected. Some of the evidence of damaging variation: almost 40% of all the work performed was at a below-average rate; the biggest problem was on Resource B, where 60% of the work was performed at a rate below 5.0.

"Variation could have occurred anywhere", said the senior superintendent. "Because we have the capacity and workload spread evenly across the processes and Resources, the variation we experience anywhere in the system is going to affect production everywhere in the system."

"What is bothersome about this arrangement?", asked the CEO. "What should cause us alarm, if RB Builders was relying on this approach? Where is the competitive separation?"

"I think the biggest potential problem would be the reliance on elevating the capacity of what are external resources", said the VP of Construction. "That capacity may not be available; those resources may not be available; they can work anywhere they want, for whomever they want; finding that capacity may very likely increase our Cost of Sales, and impact our margins.

"I am all-in for Epic Partnering®, but we cannot take the easy way out, by resorting to the tired concept of letting the answer to the question of more Throughput be more production capacity."

"The 'more for more' conundrum", said the senior superintendent.

"Those comments point to another discussion we will get to later", said the intrepid, results-based consultant. "In the homebuilding industry, the question of growth always finds its answer in more – in more-for-more; more markets, more segments, more products, more capacity, more inventory, more capital. The answer to the growth question always tends to be broad and shallow.

"I want to plant a mental image, and I want you to think about it; I want you to think about what the term 'strip-mining the value stream' connotes.

"I want you to think about that term – 'strip-mining the value stream' – in light of the fact that, in one of the Reference Point® executive management surveys we conducted, the executives we asked expected all of the growth for their homebuilding enterprises to occur as a result of some combination of three strategies: (1) geographic expansion; (2) higher market share in existing markets and buyer segments; and (3) higher market share in additional-but-already-existing buyer segments.

"Conjure it. We will come back to it."

"During the earlier versions of the game, you saw the problem that variation creates in a system with balanced production capacity", said the intrepid, results-based consultant. "It is not possible to achieve balanced flow and production in a system with balanced capacity, because we have the same capacity, and the same work requirement, at every resource.

"There is no-thing to manage, because we have to manage every-thing.

"We want predictable, even production flow, but predictable and even production flow is an outcome, not a mechanism. To achieve even-flow, we have to unbalance the system, and we have to do it by design, by intent. Actually, we have to dispel the illusion of balanced capacity and embrace what is reality, what we know to be true: systems have limitations; systems have constraints.

"Remember our analogy of a chain."

"Constraints exist, whether we want them to exist or not. If there was no constraint, a system would have unlimited capacity. Yet, we know that systems do not have unlimited capacity.

"There is always a constraint", she said. "There is always one link that is weaker than any of the other links.

"We have to accept the location of some constraints: for example, a subcontractor that is in high demand or in short supply; even though that subcontractor is an external resource, he would be a constraint internal to the system, a production constraint, as opposed to a constraint external to the system, like a market constraint. Our preference would be to determine the constraint resource ourselves, and place it where we want it, where it does the most good, as a pacemaker that pulls work into the system.

"Think about our repeated discussions about the Constraint Capacity Resource, like the subcontractor I just noted. The acknowledgement of a CCR means that we have to live with some level of excess/reserve capacity on the non-constraint, non-pacemaker resources; that's a good thing, not a bad thing.

"In a Pipeline game™, there are two ways to reflect constraints, and a system with unbalanced capacity", she said. "One way is to use Resources with different capacities; for example, we could have three 'strong' Resources and one 'smart' Resource, and let the 'smart' Resource be the constraint. The other way is to let the constraint reflect the result of improving parts of the process, reflect the result of removing waste and non-value-adding effort from a portion of the workflow.

"In Game 4B, we are going to impose a constraint by using improved Resources with different capacities: Resources A, B, and D will be 'strong' Resources, each with an average capacity of five; Resource C will be a 'smart' Resource, with an average capacity of four; all four Resources have the same probability-distribution profile, all four Resources have the same work requirement.

"Which is the constraint?"

"Resource C", said one of the sales representatives. "Probably because it's a superintendent."

"What are your projections?", asked the intrepid, results-based consultant.

"Resource C is going to define the capacity of the system, because it is the constraint", said the senior superintendent. "Resource C has a work requirement of four, and a capacity of four, so I would say we should be able to complete a project every round; we play 12-round games; 12 Projects Completed.

"There is no change in either Projects Invested or Projects Consumed, because Cost of Sales is a stipulated variable cost, and Operating Expense is a non-variable cost imposed as a budget that spans the entire game", he said. "So – 12 Projects Completed; Projects Invested is 75% of that, so Cost of Sales is nine projects; Gross Income, or Throughput, is three projects.

"Operating Expense is a budget of two Projects Consumed per game imposed without regard to the capacity of the resources, so the residual Net Income is one Net Project Completed. Net Income Margin should be 8.3%, which is also Return on Sales; that covers the margin side of economic return.

"We shouldn't need to carry more than four Projects-in-Process, because Resources A, B, and D have reserve capacity, owing to their higher work rate relative to the constraint; they can recover, Resource C cannot.

"Actually, what we need is to never leave Resource C idle", the senior superintendent said. "If we lose Throughput on that process, we lose it for the entire game; four Projects-in-Process might be the right number, but I would say we need to always have two projects/jobs in B4 at the end of each round, so that Resource C always has enough work in the next round; we need to exploit Resource C.

"If we project four Projects-in-Process as the average work-in-process, then 12 Projects Completed divided by four Projects-in-Process gives us an Inventory Turn of 3x. Cycle Time is the reciprocal of Inventory Turn, so four Projects-in-Process divided by 12 Projects Completed, times the number of periods is four periods; four periods times 30 days per period equals 120 days.

"You forgot about our operating performance metric stepchild", said the intrepid, results-based consultant. "Productivity rate should be 12 Projects Completed divided by two Projects Consumed; six completed projects per game.

"Familiar territory on the projections. Let's play it."


## GAME 4B


Game 4B was played with 'strong' Resources at the A, B, and D processes, and a 'smart' Resource at the C process. The workload across processes was balanced, but the Resource capacities were not. Resource A was still filled by a sales representative, B, C, and D still filled by superintendents. Necessary WIP was thought to be four Projects-in-Process, because of the capacity of the C Resource; no Minimum or Maximum WIP was imposed, but the team was encouraged to finish each round with two projects/jobs sitting in the B4 (B complete) task.

## GAME 4B

| | G-1B | G2B | G3B | G4B | | | | |
|---|---|---|---|---|---|---|---|---|
| Description<br>Std. = 4<br>Smart = 4<br>Strong = 5 | Balanced, 4 Std., Necessary WIP = 4 | Balanced, 4 Smart, Necessary WIP = 4 | Balanced, 4 Strong Necessary WIP = 5 | Balanced, 3 Strong, 1 Smart, WIP =4 | | | | |
| | | | | | | | | |
| **Budgeted** | | | | | | | | |
| Revenue | 12 | 12 | 15 | 12 | | | | |
| P-I-P | 4 | 4 | 5 | 4 | | | | |
| OE | 2 | 2 | 2 | 2 | | | | |
| P-Rate | 6 | 6 | 7.5 | 6 | | | | |
| Inv. Turn | 3X | 3X | 3X | 3X | | | | |
| CT (days) | 120 | 120 | 120 | 120 | | | | |
| NI | 1 | 1 | 2 | 1 | | | | |
| NI % | 8.3% | 8.3% | 13.3% | 8.3% | | | | |
| ROIA | 25% | 25% | 40% | 25% | | | | |
| | | | | | | | | |
| **Actual** | | | | | | | | |
| Revenue | 9 | 10 | 14 | 11 | | | | |
| P-I-P | 4 | 5.5 | 7.8 | 4 | | | | |
| OE | 2 | 2 | 2 | 2 | | | | |
| P-Rate | 4.5 | 5.0 | 7.0 | 5.5 | | | | |
| Inv. Turn | 2.3X | 1.8X | 2.2X | 2.8X | | | | |
| CT (days) | 160 | 198 | 201 | 131 | | | | |
| NI | .25 | .50 | 1.50 | .75 | | | | |
| NI % | 2.8% | 5.0% | 10.7% | 6.8% | | | | |
| ROIA | 6.3% | 9.0% | 23.6% | 19.0% | | | | |
| | | | | | | | | |
| Resource A | 3.6 | 4.1 | 4.9 | 5.4 | | | | |
| Resource B | 4.2 | 3.9 | 4.7 | 4.9 | | | | |
| Resource C | 3.7 | 4.1 | 5.1 | 4.2 | | | | |
| Resource D | 4.3 | 3.9 | 5.0 | 5.1 | | | | |
| | | | | | | | | |

"Go ahead", said the intrepid, results-based consultant "Make my day."

"Not quite as good as we expected", said the CEO. "There was better balance between margin and velocity. What stood-out was how manageable the system is, when you actually have something to manage. We had no difficulty controlling work-in-process; it was very straight-forward, with a simple order: keep sufficient work in front of the constraint.

"There was a single glitch, during Rounds 8 and 9, caused by back-to-back production problems on the constraint, on Resource C. We lost the Throughput on a single round, on

Round 9. That occurrence – and that occurrence alone – accounted for the lower Net Income Margin and Return on Sales; 6.8%, instead of 8.3%. "Productivity rate was 5.5 projects, instead of six projects; Inventory Turn was 2.8x, instead of 3x, all due to the lower-than-projected Throughput; Cycle Time was 131 days, instead of 120 days, again, due to the lower Throughput.

"ROIA was at 19%, instead of 25%, equal parts lower margin and less velocity, but all due to the single Project Complete that we missed."

"I want everyone to take note of the remarkable restraint that Resource A demonstrated throughout the game", said the VP of Sales. "She just happens to be on my sales team, and we are so proud of her. She rolled lights-out, the top-performing resource in the system.

"She could have made it rain, brothers and sisters, flooding the system with starts. But, she didn't; instead, Resource A did her part to support Resource C; not too many project starts, not too few."

"We're proud of you, Chief", said the CEO. "You are a model of restraint. Keep it up."

The intrepid, results-based consultant pulled up the PowerPoint slide of a production system with unbalanced capacity.

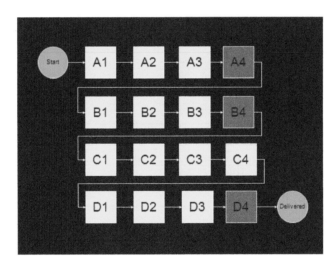

"You have seen this depiction of an unbalanced capacity system in previous versions of the game", she said. "It is a different way to show the benefit of a constraint, showing, instead, the improvement from a reengineered process. In this case, the improvement results in the A, B, and D phases having fewer tasks to perform.

"It's really the picture of a process of continuous improvement. The tasks eliminated were muda – wasted, non-value-adding effort that no one is going to miss, and now the A, B, and D phases deliver the same value – the same Throughput – with less wasted effort, less duration; those Resources have the same workflow capability they had before the improvement, so the difference between work and capacity becomes excess/reserve capacity.

"The C-Phase has the same number of tasks as before;  it is alone in that regard.  It didn't register improvement, perhaps because each of its tasks were necessary, perhaps because there wasn't as much that could be improved.  Regardless of the circumstances, the C-Phase has no reserve capacity, because its capacity is equal to the work demanded of it;  the C-Phase has become the system constraint;  and, Resource C is now the Constraint Capacity Resource, the CCR.

"It has become the system's constraint – not because we added capacity to the other phases – because we improved the process of the other phases;  we improved the process, itself.  There is a difference.

"There is a difference when we decide to acquire more production capacity.  One way – the easy way – is to find more external resources to perform the work, take a 'more-for-more' approach to solve the problem.  Another way – a better, but more difficult way – is to make the existing external resources more productive, make them able to do more with the capacity they have;  it's a product of Epic Partnering®, investing in the capabilities of our suppliers and subcontractors.

"Separate and apart from whether RB Builders chooses to acquire more resource capacity, there is the question of what type of work are we asking resources to do?  Is it non-value-adding work that we expect them to perform faster?  Or, will we reengineer the underlying process, so that resources only perform work that adds value?

"The later choice is what we see with this unbalanced capacity production system;  not only do we now have a clear constraint to manage, but we have a better process.  We have a process that delivers the same value with less waste, less rework, less redundancy, fewer mistakes, fewer delays."

"And – less frustration, less resentment", said the VP of Construction.

"For this next game – Game 5B – let's look at a scenario in which most of the effort has been focused on improving the process", she continued.  "We have made progress on driving variation out of the system, but we're not where we want to be;  we have one 'smart' Resource on the C-Phase, but the A, B, and D Resources are 'standard' Resources.

"Resource C has less variation than the other Resources, but not a higher work rate;  we have not yet partnered long enough – at an epic level – to develop 'strong' or 'premium' Resources.  That is still to come.

"What are your expectations?"

"Resources define the capacity of the system", said one of the superintendents.  "All of our Resources have the same average work rate;  therefore, they have the same capacity.  Resources A, B, and D have less work to perform, three tasks instead of four; that makes Resource C the constraint.

"Resources A, B, and D are less stable, due to the variation they can experience, even though they have the same capacity as each other and Resource C", he said. "Frankly, we have not seen this resource configuration before, with unbalanced capacity; we are going to have to figure out how to handle the instability."

"Do you recall how a system protects itself from variation?", asked the intrepid, results-based consultant.

"A system will protect itself in one of three ways", answered a sales representative. "It will insist on having some combination of additional capacity, higher levels of work-in-process, or longer durations."

"The system has all the capacity it's going to get", said the superintendent. "My guess is that higher variation – less stability – on the A, B, and D Resources is going to require more Projects-in-Process. How many? I don't have an answer. If the system does buffer itself with higher inventory, and cannot increase the number of Projects Completed, it will translate into longer duration."

The CFO pondered the discussion. "Resources A, B, and D have capacity in excess of their work requirement, but they also have a lot of variation", he said. "I think we have to base the projections off of the constraint, Resource C; four tasks, average capacity of four; we should be able to complete one project per round.

"The projected Revenue is 12 Projects Completed; Cost of Sales is stipulated at 75% of Projects Completed, which would be nine Projects Invested; Throughput – Gross Income – should be three Projects Completed; Operating Expense is an imposed budget of two Projects Consumed, somewhat of a fixed cost; the projected Net Income would be one Net Project Completed; the projected Net Income Margin, which we also use for the Return on Sales component of ROIA, would be 8.3%.

"There is the question of variability buffering", he said. "But, presuming an average of four Projects-in-Process is sufficient, and presuming we can achieve the projected Revenue of 12 Projects Completed, then we have to say the projections for a game played with 'standard' or 'smart' Resources are achievable; a Productivity rate of six projects, a 3x Inventory Turn, a 120 day Cycle Time.

"With a Return on Sales of 8.3% and an Inventory Turn of 3x, Return on Invested Assets should be 25%, rounded."

"We shall see", said the CEO.


## GAME 5B

Game 5B was played with 'standard' Resources at the A, B, and D processes, and a 'smart' Resource at the C process, in an unbalanced system. The same RB Builders team was on the game board. Necessary WIP was thought to be four Projects-in-Process, but there was no Minimum or Maximum WIP imposed.

## GAME 5B

| | G-1B | G2B | G3B | G4B | G5B | | | |
|---|---|---|---|---|---|---|---|---|
| Description<br>Std. = 4<br>Smart = 4<br>Strong = 5 | Balanced, 4 Std., Necessary WIP = 4 | Balanced, 4 Smart, Necessary WIP = 4 | Balanced, 4 Strong Necessary WIP = 5 | Balanced, 3 Strong, 1 Smart, WIP =4 | Un-Bal. 3 Std. 1 Smart WIP = 4 | | | |
| | | | | | | | | |
| Budgeted | | | | | | | | |
| Revenue | 12 | 12 | 15 | 12 | 12 | | | |
| P-I-P | 4 | 4 | 5 | 4 | 4 | | | |
| OE | 2 | 2 | 2 | 2 | 2 | | | |
| P-Rate | 6 | 6 | 7.5 | 6 | 6 | | | |
| Inv. Turn | 3x | 3x | 3x | 3x | 3x | | | |
| CT (days) | 120 | 120 | 120 | 120 | 120 | | | |
| NI | 1 | 1 | 2 | 1 | 1 | | | |
| NI % | 8.3% | 8.3% | 13.3% | 8.3% | 8.3% | | | |
| ROIA | 25% | 25% | 40% | 25% | 25% | | | |
| | | | | | | | | |
| Actual | | | | | | | | |
| Revenue | 9 | 10 | 14 | 11 | 10 | | | |
| P-I-P | 4 | 5.5 | 7.8 | 4 | 4.3 | | | |
| OE | 2 | 2 | 2 | 2 | 2 | | | |
| P-Rate | 4.5 | 5.0 | 7.0 | 5.5 | 5.0 | | | |
| Inv. Turn | 2.3x | 1.8x | 2.2x | 2.8x | 2.3x | | | |
| CT (days) | 160 | 198 | 201 | 131 | 155 | | | |
| NI | .25 | .50 | 1.50 | .75 | .50 | | | |
| NI % | 2.8% | 5.0% | 10.7% | 6.8% | 5.0% | | | |
| ROIA | 6.3% | 9.0% | 23.6% | 19.0% | 11.5% | | | |
| | | | | | | | | |
| Resource A | 3.6 | 4.1 | 4.9 | 5.4 | 3.7 | | | |
| Resource B | 4.2 | 3.9 | 4.7 | 4.9 | 4.1 | | | |
| Resource C | 3.7 | 4.1 | 5.1 | 4.2 | 4.1 | | | |
| Resource D | 4.3 | 3.9 | 5.0 | 5.1 | 3.8 | | | |
| | | | | | | | | |

"Results?", asked the intrepid, results-based consultant.

"The results in Game 5B were equal or better to either of the other two games in which we used only 'standard' or 'smart' Resources", said the CFO.  "The results were better in almost every category.

"But – that wasn't the case compared with simulations involving 'strong' Resources; the results in Game 5B were not as good. Clearly, capacity in excess of work requirements is an advantage; it might be necessary, but is it sufficient?

"This game, specifically: we fell two Projects Completed short of target; since Cost of Sales is 75% of Revenue, our Gross Income – or Throughput – was 2.5 Projects Completed; Operating Expense is an imposed budget of two Projects Consumed, which gave us Net Income – Net Projects Completed – of .5 projects. Net Income Margin – which is also Return on Sales in the ROIA calculation – was 5.0%.

"We had a Productivity rate of 5.0 projects, compared with our target rate of 6.0 projects; Inventory Turn was only 2.3x, with fault lying on the numerator, as well as the denominator – not enough Projects Completed, too many Projects-in-Process; our Cycle Time was 155 days.

"Return on Invested Assets was 11.5%, well less than half of the economic return we projected."

"It isn't the first time we have used them, but coping with the variation on 'standard' Resources was a huge challenge", said the senior superintendent. "Resource A and B both fell short of the expected average work rate of four tasks per round each; combined, they fell short, as well. As adjoining resources, A and B managed to coordinate their efforts enough to support the constraint, Resource C; plus, both of their averages were above the amount of work their processes required.

"Consequently, Resource C didn't appear starved for projects, and seemed able to give Resource D sufficient projects upon which to work, most of the time. Resource C averaged 4.1, higher than what it needed, and only had two periods in which its output was below standard. Resource D was the bigger problem; it couldn't close out the projects."

The senior superintendent thought about that last point. "Let me clarify that point. For the game, Resource D averaged 3.8, higher than its work requirement, but there were five periods – count 'em, five – in which it produced less than the work requirement.

"The performance of Resource D was devastating. There were three periods in which it delivered two projects/jobs, but five periods in which it couldn't close any projects/jobs; a 50% failure rate. We haven't seen this level of variation in Throughput in other production scenarios, in which two-thirds of the periods produced unexpected results. In this game, the amount of variation killed us.

"Improving the process is not sufficient."

"As Jack Welch always likes to remind us", the CEO said. "Variation is evil."

"Okay, so we don't think improving the process alone is the answer", said the intrepid, results-based consultant. "Let's run the next game – Game 6B – with 'smart' Resources at all four processes. Same capacity; less variation. We would expect the system to be more stable, more predictable, more manageable."

"Same system, same capacity", said the CFO. "The projected results should be the same as the previous game."

"Play it", said the intrepid, results-based consultant.

## GAME 6B

Game 6B was played with four 'smart' Resources, in an unbalanced system. The team makeup did not change. Projections were the same as the previous game (5B).

"How did you do?", said the intrepid, results-based consultant.

"Revenue was 11 Projects Completed", replied the CFO. "Gross Income was 2.75 projects, which gave us Net Income of .75 Net Projects Completed, which resulted in a Net Income Margin/Return on Sales of 6.8%. Productivity was 5.5; our Inventory Turn was 2.4x; Cycle Time was 150 days.

"All of these operating and financial performance measures were marginally better than the ones we generated in the previous game.

"Return on Invested Assets benefited from slightly better margins and slightly higher velocity; the 6.8% Return on Sales and the 2.4x Asset Turn produced an ROIA of 16.3%. Compared with the 11.5% ROIA in Game 5B, we were able to stretch the difference out a bit more on economic return.

"But – you just had a sense, we can do better."

## GAME 6B

| Description<br>Std. = 4<br>Smart = 4<br>Strong = 5 | G-1B<br>Balanced,<br>4 Std.,<br>Necessary<br>WIP = 4 | G2B<br>Balanced,<br>4 Smart,<br>Necessary<br>WIP = 4 | G3B<br>Balanced,<br>4 Strong<br>Necessary<br>WIP = 5 | G4B<br>Balanced,<br>3 Strong,<br>1 Smart,<br>WIP =4 | G5B<br>Un-Bal.<br>3 Std.<br>1 Smart<br>WIP = 4 | G6B<br>Un-Bal.<br>4 Smart<br>WIP = 4 | | |
|---|---|---|---|---|---|---|---|---|
| | | | | | | | | |
| Budgeted | | | | | | | | |
| Revenue | 12 | 12 | 15 | 12 | 12 | 12 | | |
| P-I-P | 4 | 4 | 5 | 4 | 4 | 4 | | |
| OE | 2 | 2 | 2 | 2 | 2 | 2 | | |
| P-Rate | 6 | 6 | 7.5 | 6 | 6 | 6 | | |
| Inv. Turn | 3X | 3X | 3X | 3X | 3X | 3X | | |
| CT (days) | 120 | 120 | 120 | 120 | 120 | 120 | | |
| NI | 1 | 1 | 2 | 1 | 1 | 1 | | |
| NI % | 8.3% | 8.3% | 13.3% | 8.3% | 8.3% | 8.3% | | |
| ROIA | 25% | 25% | 40% | 25% | 25% | 25% | | |
| | | | | | | | | |
| Actual | | | | | | | | |
| Revenue | 9 | 10 | 14 | 11 | 10 | 11 | | |
| P-I-P | 4 | 5.5 | 7.8 | 4 | 4.3 | 4.6 | | |
| OE | 2 | 2 | 2 | 2 | 2 | 2 | | |
| P-Rate | 4.5 | 5.0 | 7.0 | 5.5 | 5.0 | 5.5 | | |
| Inv. Turn | 2.3X | 1.8X | 2.2X | 2.8X | 2.3X | 2.4X | | |
| CT (days) | 160 | 198 | 201 | 131 | 155 | 150 | | |
| NI | .25 | .50 | 1.50 | .75 | .50 | .75 | | |
| NI % | 2.8% | 5.0% | 10.7% | 6.8% | 5.0% | 6.8% | | |
| ROIA | 6.3% | 9.0% | 23.6% | 19.0% | 11.5% | 16.3% | | |
| | | | | | | | | |
| Resource A | 3.6 | 4.1 | 4.9 | 5.4 | 3.7 | 3.8 | | |
| Resource B | 4.2 | 3.9 | 4.7 | 4.9 | 4.1 | 4.4 | | |
| Resource C | 3.7 | 4.1 | 5.1 | 4.2 | 4.1 | 3.9 | | |
| Resource D | 4.3 | 3.9 | 5.0 | 5.1 | 3.8 | 4.0 | | |
| | | | | | | | | |

"The only change between Game 5B and Game 6B was swapping out the three 'standard' Resources for three 'smart' Resources", said one of the sales representatives. "Did it make a difference in the level of variation?"

"The average work rate across all four 'smart' Resources during this game was 4.0, compared with the overall work rate of 3.9 with the mix of Resources in the previous game", said the senior superintendent. "There was an almost identical percentage of sub-substandard work rate; 35% in the previous game, 33% in this game; the percentage in the current game hit the expected distribution exactly; the high/low spread was +.2/-.3 in Game 5B and +.4/-.2 in Game 6B.

"Whereas the D Resource was the problem in the previous game, the C Resource was the problem in the current game; Resource C was expected to be the constraint, and it was. Resources A and B had no problems supplying Resource C, but Resource C struggled, with five sub-standard work periods.

"It was troubling that Projects-in-Process was higher in Game 6B than it was in Game 5B. That was not the expectation."

"After you have done everything possible to 'exploit' the system's constraint – to make certain that it is never idle, make sure it works as fully as it can – the only option left is to 'elevate' it", explained the intrepid, resulted-based consultant. "You have to increase capacity on the constraint.

"In a manufacturing environment, that's a bigger decision, to the extent capacity and costs are defined and linked by internal Resources. It is an easier decision in a homebuilding production environment, because capacity is determined by external Resources that are the direct, variable costs associated with Cost of Sales, which at least means it's not an overhead-type cost decision.

"Oh – and did I mention something about strip-mining a value stream?

"Just food for thought.

"For this next game, I want you to replace the 'smart' Resource on the C-Process with a 'strong' Resource", she said. "Through your hard work and diligent effort, you have found a way to give Resource C more capacity, without giving anything away on the margin side; consider it to be evidence of the result of all your efforts at Epic Partnering®.

"Resources A, B, and D will remain 'smart' Resources, with 33% more capacity than they need, because they operate in improved processes; comparatively-speaking, Resource C has reserve capacity of 25%; even though it has more capacity than it really needs, Resource C remains the constraint."

"It seems to me, if you increase capacity, you have to increase the amount of work-in-process, or you risk starving the constraint", said one of the superintendents. "If you increase the capacity of the CCR by 25%, do you have to increase the number of Projects-in-Process accordingly?"

"The most important consideration is ensuring full utilization of the Capacity Constraint Resource, Resource C", said the intrepid, results-based consultant. "Resources A and B have more than enough capacity to service Resource C, provided they coordinate their decisions.

"Resources A and B have to find a way to keep two Projects-in-Process in the B3 task at all times, ready to be pulled into the C-Phase. The condition of Projects-in-Process in B3 has to become the pull mechanism – the release mechanism – for Projects-Ready, which is that pile of chips sitting in front of A1.

"So – what are the projections?"

"The capacity of the constraint defines the capacity of the system", said the senior superintendent. "I'm not completely comfortable with the capacity differential between Resource C and Resources A, B, and D; reserve capacity of 33% versus 25% is not a significant difference, and we may be too close to balanced capacity. But, we have to suppose the system can produce Revenue of 15 Projects Completed."

"Then, we have the same operating statement as Game 3B", said the CFO, pointing to the comparative results scorecard. "Cost of Sales is a direct, variable cost stipulated at 73% of Projects Completed, so Throughput – Gross Income – should be four Projects Completed, rounded. Operating Expense is an indirect, non-variable cost imposed as a budget of two Projects Consumed; Net Income is a residual of two Net Projects Completed, which gives us a Net Income Margin/Return on Sales of 13.3%.

"The rate of Productivity should be 7.5 projects; if we restrict Inventory to five Projects-in-Process, Inventory Turn will be 3x; Cycle Time will be 120 days. ROIA should be ROS of 13.3% times and Asset Turn of 3x; ROIA should be 40%."

"Play it."


## GAME 7B


Game 7B was played with three 'smart' Resources (A, B, and D) and one 'strong' Resource (C), in an unbalanced system. The sales representative that had been Resource A switched roles with the superintendent that had been Resource D; projections were the same as Game 3B.

"It's about time you guys made a girl the closer", said the intrepid, results-based consultant.

"That's what we do", said the VP of Sales. "Close sales. We make it rain."

"We know, Chief", said the CEO.

## GAME 7B

| | G-1B | G2B | G3B | G4B | G5B | G6B | G7B | |
|---|---|---|---|---|---|---|---|---|
| Description<br>Std. = 4<br>Smart = 4<br>Strong = 5 | Balanced, 4 Std., Necessary WIP = 4 | Balanced, 4 Smart, Necessary WIP = 4 | Balanced, 4 Strong Necessary WIP = 5 | Balanced, 3 Strong, 1 Smart, WIP =4 | Un-Bal. 3 Std. 1 Smart WIP = 4 | Un-Bal. 4 Smart WIP = 4 | Un-Bal. 3 Smart, 1 Strong WIP = 5 | |
| | | | | | | | | |
| **Budgeted** | | | | | | | | |
| Revenue | 12 | 12 | 15 | 12 | 12 | 12 | 15 | |
| P-I-P | 4 | 4 | 5 | 4 | 4 | 4 | 5 | |
| OE | 2 | 2 | 2 | 2 | 2 | 2 | 2 | |
| P-Rate | 6 | 6 | 7.5 | 6 | 6 | 6 | 7.5 | |
| Inv. Turn | 3X | 3X | 3X | 3X | 3X | 3X | 3X | |
| CT (days) | 120 | 120 | 120 | 120 | 120 | 120 | 120 | |
| NI | 1 | 1 | 2 | 1 | 1 | 1 | 2 | |
| NI % | 8.3% | 8.3% | 13.3% | 8.3% | 8.3% | 8.3% | 13.3% | |
| ROIA | 25% | 25% | 40% | 25% | 25% | 25% | 40% | |
| | | | | | | | | |
| **Actual** | | | | | | | | |
| Revenue | 9 | 10 | 14 | 11 | 10 | 11 | 15 | |
| P-I-P | 4 | 5.5 | 7.8 | 4 | 4.3 | 4.6 | 4.9 | |
| OE | 2 | 2 | 2 | 2 | 2 | 2 | 2 | |
| P-Rate | 4.5 | 5.0 | 7.0 | 5.5 | 5.0 | 5.5 | 7.5 | |
| Inv. Turn | 2.3X | 1.8X | 2.2X | 2.8X | 2.3X | 2.4X | 3.1X | |
| CT (days) | 160 | 198 | 201 | 131 | 155 | 150 | 118 | |
| NI | .25 | .50 | 1.50 | .75 | .50 | .75 | 2 | |
| NI % | 2.8% | 5.0% | 10.7% | 6.8% | 5.0% | 6.8% | 13.3% | |
| ROIA | 6.3% | 9.0% | 23.6% | 19.0% | 11.5% | 16.3% | 41.2% | |
| | | | | | | | | |
| Resource A | 3.6 | 4.1 | 4.9 | 5.4 | 3.7 | 3.8 | 4.1 | |
| Resource B | 4.2 | 3.9 | 4.7 | 4.9 | 4.1 | 4.4 | 3.8 | |
| Resource C | 3.7 | 4.1 | 5.1 | 4.2 | 4.1 | 3.9 | 4.8 | |
| Resource D | 4.3 | 3.9 | 5.0 | 5.1 | 3.8 | 4.0 | 4.2 | |
| | | | | | | | | |

"Results?", asked the intrepid, results-based consultant.

"We hit or exceeded all of our numbers", replied the CEO. "The real story was work-in-process. Our objective was to keep our average Inventory – our average work-in-process – at five Projects-in-Process. During the game, it ranged between four and six Projects-in-Process, but the ending Inventory in each round was at five Projects-in-Process or less, more than 90% of the time.

"We controlled our Inventory that well, despite the A and B Resources not quite holding their expected work rate. That was probably the biggest challenge, our two least-productive Resources having to supply our constraint with enough work."

"What did you expect? They're superintendents", said the VP of Sales.

"It's the bottom of the 12th . . . down a Project Complete . . . projects on C4 and D1 . . . last chance to make budget . . . Resource D steps to the plate . . . she been on fire . . . needs to produce at least five tasks, right now, though . . .

. . . there it is . . . walk-off five."

"Chief's cleverness aside, this game is a picture of how thin the margin of success is in the homebuilding business", said one of the superintendents. "And – how consistent we have to be, period by period.

"In effect, you have to get the job done, every day."

The intrepid, results-based consultant surveyed the faces in the conference room.

"It's a blend of improvements to consider: improved processes; more productive resources; reducing variation; shorter schedules, faster turns. It's about a disciplined approach – a planned, finite, controlled approach – to managing work-in-process; it's about maximizing the rate of throughput, every period.

"It's not about the parts", she said. "It's about the system. It's not about treating the symptoms of problems; it's about solving their root cause. It's not about balanced capacity at every resource; it's about managing the constraint."

"So – what is this mental image of 'strip-mining the value stream' you have been mentioning?", asked the VP of Construction.

The senior superintendent spoke. "I think this is what she means: homebuilding is an industry that outsources everything; it outsources everything 'above the line', all of the work reflected in our Cost of Sales, is done by someone other than RB Builders; except for land, everything we sell is produced by our subcontractors or furnished by our suppliers.

"What she is saying is that all of the Gross Income we generate – all of the Throughput, all of the Contribution that would be available to absorb overhead – is diminished before it gets to us.

"That's what she's been saying: we strip-mine the value stream; almost all of the benefit from creating value for the buyer goes to someone else. As we know excruciatingly well, homebuilding is not a high margin business; that knowledge has been reinforced in the Pipeline games™. As a result, we have to look to geographic expansion, market share, or segment share as the mechanisms for making more money. So – we get locked into a more-for-more proposition."

"What about Epic Partnering®?", asked the VP of Sales. "Listening to you guys, it's not like suppliers and subcontractors are our adversaries. Without them, there is no way to create the value that my team sells – and sells so well, I might add."

"Adversaries? No, not at all. And – yes – we are making strides on the trade partner front", replied the CEO. "But, we also see how incredibly fragmented and difficult-to-manage the value stream is in this industry, especially compared to most manufacturing industries. We acknowledge the significant amount of additional coordination it requires.

"Job schedules are not the most complicated workflow structures on the planet, but – in terms of duration and number of tasks – I would put the complexity of the work breakdown structure for project portfolio management up against the workflow structure of any manufacturing process.

"Our trade partners are all profitable businesses, due in no small part to our efforts to partner with them, and their willingness to partner with us. Yet, all of us have separate overheads that we must absorb, a requirement made all the more difficult by the complexity of fragmentation.

"We also are painfully aware of the labor difficulties confronting the construction industries, as a whole. It is a persistent shortage, and it is structural. It is not going to go away anytime soon, and we have to compete for the constrained services of suppliers and subcontractors.

"Where were you going with this, anyway?", he asked, turning to the intrepid, results-based consultant.

"The same place you're going with it", she said.

"Before you go anywhere with it, I have a question", said one of the superintendents. "What is the difference between a 'value stream' and a 'supply chain'? At times, we seem to use those terms interchangeably."

"They are similar but different, but more similar than different", said the intrepid, results-based consultant. "You're right, the terms 'supply chain' and 'value stream' are often used interchangeably; 'supply chain' is used more in logistics and is the broader of the terms, while 'value stream' is a Lean Production term; value streams emphasize value, what creates it; they get mapped, so they tend to be more visual, and they talk more about current and future states; supply chains and value streams both deal with information, materials, vendors, customers; they can be either, but supply chains tend to be more inter-company, value streams more intra-company.

"In theological terms, they would be considered different denominations of the same religion.

"Whether you call it supply chain management or value stream mapping, we are dealing with the coordination of activities, information, and materials of a house, from start to completion.

For the purposes of managing homebuilding production, in my opinion, there are no meaningful distinctions between the two terms."

The intrepid, results-based consultant turned her attention back to the rest of the team.

"The discussion about strip-mining the value stream is about whether vertical integration is an alternative strategy to traditional growth", she continued. "I don't know that this is the right economy and housing market for RB Builders to consider trying to vertically integrate its supply chain. It would be quite an undertaking.

"We talk about the need to build urgency towards results. In all likelihood, there are more important, more immediate concerns that require your attention, more important, more immediate benefits available to you.

"Perhaps we need to be more tactical than strategic, right now; perhaps RB Builders should focus, for now, on vertical integration as a more effective method of managing its Constraint Capacity Resource, and leave it there.

"What we can do right now, however, is have the Pipeline game™ reflect a vertically-integrated homebuilding company", she said, moving towards the erasable board. "I think we can at least demonstrate what is at stake."

"Let's do it", said the CEO.

"Start by giving me the breakdown of Cost of Sales, all of the direct, variable costs of building a house", said the intrepid, results-based consultant. "And – remember – you have to use a variable costing approach that treats sales commissions and the financing costs under the construction lines of credit as direct, variable costs that need to be included in Cost of Sales.

"What percentage of Cost of Sales is attributable to land? What percentage can be attributed to commissions and financing? Of the remainder, what is the breakdown between labor and material?"

The CEO reached over, placed the conference room phone on speaker, and entered an extension number. When the voice answered, the CEO said, "Could you please come in here?"

Less than a minute later, RB Builders' Purchasing Manager walked into the conference room. The CEO introduced him to the intrepid, results-based consultant and motioned him to take a seat. Then, the CEO turned to the CFO.

"Land, commissions, and financing?", he asked. "What is the breakdown on costs, generally."

"I presume you are talking about developed lots", the CFO said. "But, it doesn't matter, because we no longer keep any of the land or developed lot inventory on RB Builders'

Balance Sheet.  The only land cost we have anywhere in the company is job costs committed on individual houses.

"Our all-in lot costs on a typical job budget is about 30% of Cost of Sales;  commissions and financing total about 6%;  the remainder – 64% – is labor and material.  These are percentages of Cost of Sales, not percentages of sales price or Revenue.

"What's going on?  You act like you don't know this stuff."

"I want to hear it from you."  The CEO turned his attention to the Purchasing Manager.  "What is the breakdown between labor and material?"

"Your question presumes we bid labor and material separately", said the Purchasing Manager.  "We are testing BIM;  right now, our CAD program produces takeoffs that we do use for unit pricing, but – even now – some of our subcontractors' pricing is still being returned as job bids, and those bids aren't broken-down between labor and material.

"I am a proponent of unit pricing, but that doesn't mean every bid comes back the way we want it.  My sense is that the breakdown is 65/35, materials to labor."

"Okay, so this is what I am hearing", said the intrepid, results-based consultant, writing on the erasable board.  "Profile of Land-Sales/Financing-Materials-Labor, as percentages of Cost of Sales:  30-6-42-22.  Right?

"But, just to make it easier to remember, let's make the breakdown 30-5-40-25.  For now.  Okay?"

*Land-Sales/Financing-Materials-Labor Profile (% of COS)*
*30-6-42-22=100%*
*30-5-40-25=100%*

"So, where are your Gross Margins these days?", she asked.  "Since my firm's days of participating in RB Builders' team-based performance compensation plan ended, I'm sure I have lost track of the real numbers."

"Margins have recovered somewhat since the dark days you helped us through", replied the CFO.  "Gross Margins are averaging 25%;  not as robust as during the Age of Homebuilder Entitlement, but better than they were when you first came."

The intrepid, results-based consultant reached across the conference table for her HP-12C and tossed it to one of the superintendents.

"I can't believe you still have this coelacanth", said the superintendent.  "RPN and everything.  What do you want to know?"

"It's still the best superintendent slayer on the market", she replied.  "I want to know labor cost – listen to me – as a percentage of Revenue."

The superintendent keyed in the numbers.

"It looks like about 19 cents out of every dollar of Revenue goes to subcontract labor", he answered. "That's in keeping with the sense that subcontract labor accounts for about 25 cents out of every dollar of Cost of Sales, or about 25 cents out of every dollar of direct variable cost."

"Wonderful", said the intrepid, results-based consultant, turning her attention back to the Purchasing Manager. "Let's break the L&M cost down further. Talk to us about how labor and material costs are marked up."

"Do you want both suppliers and subcontractors?", asked the Purchasing Manager.

"No", she replied. "We just need to reflect what our subcontractors do with their costs, primarily their labor costs. When our subcontractors are supplying material under a job bid arrangement, we have to figure out how to leave the cost of that material in Cost of Sales, without their markup. The material provided by our suppliers is not a concern, because that arrangement is not going to change as a result of any type of vertical integration.

"This is a Pipeline game™; it's a sacred, mythical experience to all of you, I know, but it just has to be a representative picture, one that helps us learn, so we operate better in the real world."

"Tell me if I am correct", said the CFO. "In the Pipeline game™, we are going to reflect situations that require a new building model and a new operating statement. We are going to unmark the cost of subcontract labor, and move it into Operating Expense; we are going to leave everything else in Cost of Sales – Land, Commissions, Financing, and Materials, everything but the unmarked Labor component.

"Correct?"

"Yes", said the intrepid, results-based consultant.

"Let's simplify this", said the Purchasing Manager. "As an overall reflection of building operations, I am comfortable with Labor at 19% of Revenue, and at 25% of Cost of Sales.

"As everyone knows, we have a labor shortage; the markup percentage varies significantly from subcontractor to subcontractor, but the best evidence is that labor is currently being marked-up an average of 25%. If we agree the average markup is 25%, the reciprocal of 1.25 is .8, so that makes the unmarked Labor cost about 15% of Revenue and 20% of Cost of Sales."

"You have to realize, Labor is going to behave differently now", said the CFO. "Labor is no longer going to be a direct, variable cost; it is going to be a direct, non-variable cost, which is a cost we have never contended with before. Therefore, identifying point-in-time is important."

"So, we need to move to percentage of Revenue", said the intrepid, results-based consultant, retrieving her HP-12C from the superintendent and returning to the erasable board to continue the list. "Instead of a profile reflecting Cost of Sales, we need a profile reflecting Revenue."

Land-Sales/Financing-Materials-Labor Profile (% of COS)
30-6-42-22=100%
30-5-40-25=100%
Land + Sales/Financing + Material + Labor (% of Revenue)
23+4+29+19=75%
Land + Sales/Financing + Material + Labor (% of Revenue)
23+4+29=56%
25+5+30=60%

"I think 25+5+30=60% would be easier to remember, like the earlier profile", she thought to herself, before striking thorough her last line. "The unit of measure in the Pipeline game™ is projects, a simple metric that requires rounding, which I think simplifies things enough; I don't think we need to simplify the percentage profile . . . "

"Would you care to think out loud?", asked the CEO.

### 1. Strong Resources – Current Performance

| Line Item | Outsourced | | Integrated | |
|---|---|---|---|---|
| Revenue | 100% | 15 | 100% | 15 |
| Cost of Sales | -73% | -11 | -56% | -8 |
| Gross Margin | 27% | 4 | 44% | 7 |
| Operating Expense | -14% | -2 | -29% | -4 |
| Net Income | 13% | 2 | 15% | 3 |

"This is a comparison between operating statements that we can use in a game played with 'strong' Resources, to demonstrate outsourced versus integrated models on the labor cost", said the intrepid, results-based consultant, as she added a title, took one column from an existing table, and added a new column.

"Like I said, the Pipeline game™ is designed to illustrate production principles. But – it is still a production simulator and business game; it has to make sense. It is not possible to match the whole project count to a specific percentage. This as close as we can get, and still have an operating statement that reflects the cost profiles we have been discussing.

"The best approach is to use percentages that reflect the cost discussion, and then round the project count.

"The new integrated building model has a lower Cost of Sales and a higher Operating Expense, because it no longer outsources subcontract labor;  its unmarked labor resources are now internal resources, instead of external resources.

"In every other respect, the two building models are the same.

"They have the same Revenue, the same amount of work-in-process, the same number of closings, and the same credit facility;  the higher Net Income Margin enjoyed by the integrated building model is attributable solely to the difference between (1) what was formerly Cost of Sales marked up as subcontract labor, and (2) what is now RB Builders' Operating Expense that is not marked up.  It presents a quite different perspective on managing variable costs versus non-variable costs.

"And – someone tell me – what is that perspective?"

"The outsourced building model makes less on every house, but it has a bigger multiplier", replied the CFO, taking over the erasable board.  "The integrated building model makes more on every house, but it has a smaller multiplier.

"It is really all about using a variable costing approach.

"Regardless of cost object – in other words, whether they are considered direct costs or indirect costs – variable costs increase in proportion to activity and remain constant per unit of activity, while non-variable costs are constant through a range of activities and decrease per unit as activity increases.

"As a percentage of Revenue, when Revenue increases, Cost of Sales stays the same, because it is a variable cost;  as a percentage of Revenue, when Revenue increases, Operating Expense decreases, because it is a non-variable cost."

"It's interesting", said one of the sales representatives.  "My first instinct was to conclude that the integrated operation would have an increased emphasis on leveraging its non-variable costs."

"What's wrong with that logic?", asked a superintendent.

"It's a double-edged sword", said the sales representative.  "From the integrated operation's perspective, failing to leverage its non-variable costs presents a higher risk, but the higher percentage its non-variable costs also means that it cannot be as big a multiplier.  Gains in leverage by the outsourced operation are a 2.0x multiplier on its Gross Margin, whereas gains in leverage by the integrated operation are a 1.4x multiplier on its Gross Margin.

"Then, there is the flipside:  the outsourced operation has a lower Gross Margin to leverage, which diminishes the benefit;  the integrated operation has a higher Gross Margin to leverage, which increases the benefit."

"The *objective* is always going to be simultaneously twofold:  exploit variable costs – that is, extract more value from them – and leverage non-variable costs;  the *emphasis*, however, will always depend on the mix", said the intrepid, results-based consultant, reclaiming the erasable board and adding another table.

"The standard formula for calculating Productivity – in virtually every industry – is Revenue divided by Operating Expense. The outsourced building operation currently has a productivity rate of 8.3, more than twice the current productivity rate of the integrated building operation at 3.7, despite the integrated building operation having almost a 50% higher Net Income Margin."

"Productivity. Interesting measure, in this case. I see why you call that measure our operating performance metric stepchild", said the VP of Construction.

"Let's just say it's a worthwhile operational measure better confined to consideration of performance improvement in a single building model, not necessarily an indicator of which model to choose", said the intrepid, results-based consultant.

"Yep, comparing apples to apples", said another superintendent.

"Yep, sheer profundity", said a sales representative.

"So – what happens if both building models were to become more productive?", asked the intrepid, results-based consultant.

"You mean, if both the operations for both the building models become significantly more productive?", asked the CEO. "Good question."

The senior superintendent motioned to the intrepid, results-based consultant to hand over the HP-12C on his way to the front of the conference room. He quickly keyed in a series of calculations, and then recorded the results into a new version of the table comparing the outsourced building model to the integrated building model.

### 2. Super Strong Resources – Performance @ 140%

| Line Item | Outsourced | | Integrated | |
|---|---|---|---|---|
| Revenue | 100% | 21 | 100% | 21 |
| Cost of Sales | -76% | -16 | -57% | -12 |
| Gross Margin | 24% | 5 | 43% | 9 |
| Operating Expense | -10% | -2 | -19% | -4 |
| Net Income | 14% | 3 | 24% | 5 |

"The operating statement showing current productivity was based on the use of 'strong' Resources, which means that the higher productivity rates require a new resource category that is even more productive yet; call them 'super strong' Resources. The way a Pipeline game™ is setup, percentages that produce even calculations are important, so let's use a resource that is 40% more productive.

"Both of these new operations have higher productivity rates; nevertheless, the relative Productivity remains the same. The outsourced operation still has twice the productivity rate of the integrated operation, 10.5 versus 5.3.

"However, the integrated operation generates two-thirds more Net Income, compared with 50% more Net Income generated in the outsourced building model.

"The performance gap is growing."

"This is a good reminder about not confusing the drivers of operating performance with the business outcomes those drivers are supposed to produce", said the intrepid, results-based consultant. "RB Builders is in business to make money, not become more productive."

"Changing the subject", said the senior superintendent. "What do the breakeven points look like with these building models? That lesson has stuck with me from the time you first started working with us.

"You tell us."

The senior superintendent made two more quick calculations, and then wrote as he spoke. "There are rounding discrepancies, but both operations achieve breakeven at more-or-less the same point, Revenue-wise, in both their current and their more-productive states.

"The outsourced operation does reach breakeven a bit quicker than the integrated operation does. If neither Operating Expense nor Gross Margin changes, achieving higher productivity will not change the breakeven order, or the gap in time; less overhead would seem to slightly trump higher Gross Margin."

*Operating Expense ÷ GM Ratio = Breakeven*
*Outsourced current: 2 Projects Consumed ÷ .27 = 7 Projects Completed*
*Outsourced new: 2 Projects Consumed ÷ .24 = 8 Projects Completed*
*Integrated current: 4 Projects Completed ÷ .44 = 9 Projects Completed*
*Integrated new: 4 Projects Completed ÷ .43 = 9 Projects Completed*

"Not as significant as I thought it might be", he said. "If they become more productive, both operations just get to their respective breakeven points a bit earlier in the calendar."

"Still, not a trivial point", said the CFO.

"We have also been presuming that an increase in productivity has no negative impact on Gross Margin", said the VP of Sales. "Yes, it is more Revenue – more sales and closings – with the same capacity: same overhead, same level of work-in-process, same amount of financing. But, is it reasonable to conclude that there is no price elasticity of supply and demand in play?"

"Does RB Builders want a wider footprint, or does it want a heavier footprint?", asked the intrepid, results-based consultant.

"That depends on what 'wanting a wider or heavier footprint' means", said the VP of Sales.

"A wider footprint implies strip-mining a larger value stream – more communities, probably more of some other stuff, too", said the intrepid, results-based consultant. "A heavier footprint implies extracting more of the value from a smaller value stream – fewer communities, probably less of some other stuff, too."

"What's with probably?", said the CEO.

"Well, price elasticity of supply and demand is going to be reflected in reduced sales prices, correct?", asked a sales representative. "So, we will have less Revenue; Cost of Sales will stay the same; Gross Margin will be reduced; Operating Expense will not be affected; Net Income will be lower. Correct?"

"Let's say the additional closings from the 40% increase in productivity resulted in an overall Revenue reduction of five percent, meaning sales prices overall had to drop five percent", said the VP of Sales.

"But, that wouldn't actually happen", said the sales representative. "The elasticity would occur on the sales prices of the incremental closings – and be reflected in the Revenue they generated. So, an overall Revenue reduction of five percent would mean that the incremental contracts and closings would incur a 15% reduction in sales prices."

Nodding her agreement, the intrepid, results-based consultant, produced another table, similar to the first two tables.

### 3. Super Strong – 140%; 5% overall Price Elasticity
#### (15% SP reduction on incremental sales)

| Line Item | Outsourced | | Integrated | |
|---|---|---|---|---|
| Revenue | 100% | 20 | 100% | 20 |
| Cost of Sales | -80% | -16 | -60% | -12 |
| Gross Margin | 20% | 4 | 40% | 8 |
| Operating Expense | -10% | -2 | -20% | -4 |
| Net Income | 10% | 2 | 20% | 4 |

"I do like the look of these operating statements", she said. "It's simple; it's easy to make the distinction between building models. It's certainly interesting conjecture; however, in the end, it all comes down to how you manage variation and uncertainty in a production system.

"It looks like we have six new production scenarios.

"Let's agree that a 40% increase in productivity would likely induce price elasticity of supply and demand, and, therefore, let's also agree that price elasticity refutes the first comparison between a more productive outsourced building model and a more productive integrated building model.

"That being the case, there are really only four new production scenarios for us to consider: (1) a baseline outsourced building model, (2) a baseline integrated building model, (3) a 40% more productive outsourced building model subject to price elasticity of supply, and (4) a 40% more productive integrated building model subject to price elasticity of supply. In other words, we include both the building models in Tables 1 and 3; we eliminate both the building models in Table 2."

Surveying the conference room, the intrepid, results-based consultant performed a quick headcount, and then created a new list on the flipchart.

Team 1: Baseline Outsourced Building Model; Base-Out or B-OBM

Team 2: Baseline Integrated Building Model; Base-In or B-IBM

Team 3: Productive Outsourced Building Model; Pro-Out or P-OBM

Team 4: Productive Integrated Building Model; Pro-In or P-IBM

"We have sufficient resources and game boards to do all four of the surviving production scenarios at the same time", she said. "Team 1 is the *Baseline Outsourced Building Model*, abbreviate it Base-Out or B-OBM; Team 2 is the *Baseline Integrated Building Model,* the Base-In or B-IBM, for short; Team 3 is the *Productive Outsourced Building Model*, abbreviate it Pro-Out or P-OBM; and Team 4 is the *Productive Integrated Building Model*, the Pro-In or P-IBM, for short.

"Select your teams, while I make some adjustments to the scorecard; remember, four Resources, one scorekeeper, per team."

"You do know, of course, that Kiper rates me the first draft pick overall", said the VP of Sales.

"We know, Chief", said the CEO. "Don't spend your signing bonus all in one place."

After the teams had been assembled, the intrepid, results-based consultant explained how the multiple, simultaneous simulations were going to work, outlining the requirements on the erasable board.

"We need a new category of resources, one we will call 'super strong' Resources", she said. "These 'super strong' Resources only appear in the two more productive building models, the P-OBM run by Team 3 and the P-IBM run by Team 4; the B-OBM run by Team 1 and the B-IBM run by Team 2 use 'strong' Resources. As you know, 'strong' Resources have an average capacity of five tasks per round; a 'super strong' Resource is 40% more productive than a 'strong' Resource, so we are looking at a resource with an average capacity of seven tasks per round, instead of five."

5 tasks x 140% = 7 tasks

"The new 'super strong' Resources have the more predictable work rate we produced with the 'smart' Resources and 'strong' Resources – which is a 33% probability distribution. They have less variation.

"We will keep the easy-recall connection between dice rolls and resource capacity intact, as best we can. This time, it relates to sums: two plus four equals *six*; one plus six equals *seven*; three plus five equals *eight*."

'strong' Resource rolls: avg. = 5; 3 or 4 = 4; 2 or 5 = 5; 1 or 6 = 6

'super strong' Resource rolls: avg. = 7; 2 or 4 = 6; 1 or 6 = 7; 3 or 5 = 8

"We want to continue to unbalance the capacity of the system in favor of a constraint, because we know that setup – a production system unbalanced in favor of a planned constraint – is a more manageable system.

"In a Pipeline game™, the constraint can be either a resource constraint or a process constraint; there will either be one resource that has less capacity than the other three resources, or there will be one process – A, B, C, or D – that has more work to do, or more tasks to perform, than the other three processes."

"It's a little tricky, but I would suggest we use a process constraint", said the VP of Construction. "That would mean that Phases A, B, and D have three tasks each to perform, and Phase C has four tasks to perform.

"Because it has more work to do than the other three resources, Resource C would be the Capacity Constraint Resource, the CCR; that also would mean that Resource C defines the capacity of the system.

"In the case of the P-OBM and the P-IBM run by Team 3 and Team 4, that capacity is 40% higher than the previous system; with those teams and building models, Resources A, B, and D have significant reserve capacity that we will have to figure out how to manage without letting WIP get out of hand."

"You realize, we could have made any of the processes the constraint", said the intrepid, results-based consultant. "There are different lessons and insights from constraints located upstream or downstream. Phase C? Does anyone think otherwise?"

Not sensing any disagreement, she continued.

"Very good. I need each team to project their performance, based on either the first or third operating statement comparisons we produced, according to the building model your team will be using.

"Game 8B is going to turn out to be four-games-in-one; four teams, all running their production scenarios at the same time. We are only going to do this once, so no mulligans.

"Some measures are specified or defined, designed into the system: measures like resource capacity, Revenue, Cost of Sales, Operating Expense; other measures, like residual Net Income and Net Income Margin, are results determinable from the specified measures.

"The level of WIP you carry, however, is up to you; your discretion, your call, you have to make the operating decisions. You will have to apply the principles you have learned; you will have to be prepared to defend your decisions."

As every team reported their calculations, she logged their projections into the top half of scorecard.

## GAMES 8B 1-4

| | G8B-1 | G8B-2 | G8B-3 | G8B-4 |
|---|---|---|---|---|
| Description<br>Standard = 4<br>Smart = 4<br>Strong = 5<br>Super Strong = 7 | B-OBM<br>B-line OS<br>unbalanced<br>A,B,C,D=<br>Strong<br>full price<br>WIP: No<br>Min/Max | B-IBM<br>B-line INTG<br>unbalanced<br>A,B,C,D=<br>Strong<br>full price<br>WIP: No<br>Min/Max | P-OBM<br>140% OS<br>unbalanced<br>A,B,C,D=<br>Super<br>Strong<br>15% price<br>elasticity<br>WIP: No<br>Min/Max | P-IBM<br>140% INTG<br>unbalanced<br>A,B,C,D=<br>Super<br>Strong<br>15% price<br>elasticity<br>WIP: No<br>Min/Max |
| Budgeted | | | | |
| Revenue | 15 | 15 | 20 | 20 |
| P-I-P (avg.) | 5 | 6 | 7 | 6 |
| OE | 2 | 4 | 2 | 4 |
| P-Rate | 7.5 | 3.8 | 10.0 | 5.0 |
| Inv. Turn | 3.0X | 2.5X | 2.9X | 3.3X |
| CT (days) | 120 | 144 | 126 | 108 |
| NI | 2 | 3 | 2 | 4 |
| NI % | 13.3% | 20.0% | 10.0% | 20.0% |
| ROIA | 40% | 50% | 29% | 66% |

"From the operating statements, we knew there would be differences in the projected performance of the two versions of two different building models, based simply on what the models required", observed the intrepid, results-based consultant. "I want each team to explain their thinking on the level of work-in-process."

She went through the rotation.

"Team 1: Baseline Outsourced Building Model. What was your thinking on wanting to maintain an average WIP of five Projects-in-Process?"

"We went with the idea of Necessary WIP, kind of a capacity-driven model", said one of the superintendents. "We have four resources, each with an average capacity of five; it seems like we would need to have at least as much work-in-process as we have capacity."

"Okay. Team 2: Baseline Integrated Building Model. Your thinking on keeping six Projects-in-Process as your WIP?"

"Same thinking as Team 1, but we wanted a small buffer", said the VP of Sales. "A system will protect itself from variation with some combination of longer duration, additional inventory, or reserve capacity; resource capacity is a given, so we decided to trigger the buffer with more work-in-process; that decision results in a projection of longer cycle time, so we're not sure which it is, but we have it."

"Interesting. Team 3: Productive Outsourced Building Model. What is your reasoning on keeping your WIP at seven Projects-in-Process?"

"Like the B-OBM group, we went with Necessary WIP for a system that is supposed to be more productive than the baseline models", said the CFO. "Same thinking; four resources, but each with an average capacity of seven tasks per round; the unsettling part is that the work requirements are three and four tasks per round. It is clearly Necessary WIP, but we are not sure if it is Sufficient WIP."

"Good distinction, necessary, but not sufficient. Team 4: Productive Integrated Building Model. Why did you choose six Projects-in-Process as your WIP?"

"According to our way of thinking – according to our mental model – necessary is more than sufficient", said one of the sales representatives. "We just think we can do more with less; we think we can do more with less than what conventional wisdom would mandate, do more with less than what a balanced capacity mental model would suggest."

"Rebels", mused the senior superintendent. "Dangerous people."

Nodding to the CFO, the intrepid, results-based consultant reminded everyone, "Although these versions of the game will present less of the problem, remember that you can round your results, state them based on fractional results, or you can simply give them a range that reflects both."

## GAMES 8B 1-4

Games 8B 1-4 were played as four simultaneous games with different building models. Team 1 used a Baseline Outsourced Building Model; Team 2 used a Baseline Integrated Building Mode; Team 3 using a Productive Outsourced Building Model; Team 4 using a Productive Integrated building Model.

## GAME 8B

| Description<br>Standard = 4<br>Smart = 4<br>Strong = 5<br>Super Strong = 7 | G8B-1 | G8B-2 | G8B-3 | G8B-4 |
|---|---|---|---|---|
| | B-OBM<br>B-line OS<br>unbalanced<br>A,B,C,D=<br>Strong<br>full price<br>WIP: No<br>Min/Max | B-IBM<br>B-line INTG<br>unbalanced<br>A,B,C,D=<br>Strong<br>full price<br>WIP: No<br>Min/Max | P-OBM<br>140% OS<br>unbalanced<br>A,B,C,D=<br>Super<br>Strong<br>15% price<br>elasticity<br>WIP: No<br>Min/Max | P-IBM<br>140% INTG<br>unbalanced<br>A,B,C,D=<br>Super<br>Strong<br>15% price<br>elasticity<br>WIP: No<br>Min/Max |
| **Budgeted** | | | | |
| Revenue | 15 | 15 | 20 | 20 |
| P-I-P (avg.) | 5 | 6 | 7 | 6 |
| OE | 2 | 4 | 2 | 4 |
| P-Rate | 7.5 | 3.8 | 10.0 | 5.0 |
| Inv. Turn | 3.0x | 2.5x | 2.9x | 3.3x |
| CT (days) | 120 | 144 | 126 | 108 |
| NI | 2 | 3 | 2 | 4 |
| NI % | 13.3% | 20.0% | 10.0% | 20.0% |
| ROIA | 40% | 50% | 29% | 66% |
| | | | | |
| **Actual** | | | | |
| Revenue | 15 | 16 | 22 | 22 |
| P-I-P (avg.) | 5 | 6 | 7 | 6 |
| OE | 2 | 4 | 2 | 4 |
| P-Rate | 7.5 | 4.0 | 11.0 | 5.5 |
| Inv. Turn | 3.0x | 2.7x | 3.1x | 3.7x |
| CT (days) | 120 | 135 | 115 | 98 |
| NI | 2 | 3 | 2.0-2.4 | 4.8-5.0 |
| NI % | 13.3% | 18.8% | 9.1% to<br>10.9% | 21.8% to<br>22.7% |
| ROIA | 40% | 51% | 28-34% | 80-84% |
| | | | | |
| Resource A | 4.7 | 5.3 | 7.3 | |
| Resource B | 4.9 | 4.5 | 7.0 | |
| Resource C | 4.9 | 4.7 | 6.9 | |
| Resource D | 5.3 | 4.8 | 7.2 | |
| | | | | |

The intrepid, results-based consultant recorded the results for each game as the teams reported them. "Team 1, you had the Baseline Outsourced Building Model. Explain your results and what you learned."

"We hit our budgets right on the nose, every metric", said the superintendent. "We hit the operating performance numbers, and the business outcomes. That was an unexpected outcome, given our previous experience with achieving projections in these games.

"The most interesting discovery was how easy an unbalanced system with sufficient resource capacity made it to manage the system at the level of work-in-process we agreed to carry. It was a very natural 'pull' system; you couldn't increase WIP by starting projects/jobs without completing earlier projects/jobs first.

"Without the discipline imposed by having to maintain a planned, finite, and controlled level of work-in-process, it would have been very easy to exceed Necessary WIP; and, previously, that is exactly what we would have done; we would have pushed starts into the system, without regard to the rate of completions.

"Resources A, B, and C all performed slightly below expectations, and Process C was the system constraint. It had to hurt our performance. Resource D performed above expectations, so it was sometimes able to make up lost ground; but, it can only work on what Resource C had completed, and what was already in its own process.

"The outsourced model is what we are accustomed to managing, so – in that regard – we don't have any particular insights to offer."

"How long have we been playing the Pipeline game™?", asked the intrepid, results-based consultant. "Eight years now?

"Unlike most of the attendees at a Pipeline workshop™, RB Builders has never attempted to 'game the system', by loading WIP early in a game, and not loading it late. But, it bears mentioning; manipulating the system is a natural temptation, one that is a response to a production scenario like a Pipeline game™ that has to have predetermined starting and ending points.

"We usually specify the appropriate levels of work-in-process, what we refer to as Necessary WIP, Maximum WIP, and Minimum WIP. But, there are times – like what we are doing now – when we give teams far more latitude in determining what level of work-in-process they carry. It is a calculated tradeoff, a decision to forgo the enforcement of production principles to allow adverse discovery.

"In a real business scenario, production occurs without regard to arbitrary beginning and ending points. So, any decision to load work-in-process early in a period and not load it late would have a devastating impact on real operating performance and economic return, before and after.

"Don't succumb to that tendency.

"Team 2, you had the Baseline Integrated Building Model. What were your results? What did you learn?"

"Team 2 had the same resource profile as Team 1, but we had the baseline integrated model, not the baseline outsourced model", said the VP of Sales. "Besides the differences in the operating statements, we had also decided to carry slightly higher work-in-process: six Projects-in-Process, versus five Projects-in-Process.

"It's interesting. We knew from the discussion leading into Game 8B that integrated building models project lower productivity rates than outsourced building models, because of the higher level of Operating Expense in relation to Revenue. The same is true of the other operating measures, Inventory Turn and Cycle Time, although for a different reason: higher levels of work-in-process.

"In any event, we beat operating performance projections in all three measures. We had projected a productivity rate of 3.8; we generated a 4.0 rate. We had projected an Inventory Turn of 2.5x; we turned inventory 2.7x. Since they are reciprocals, it would stand to reason that we would also have shorter duration; we had a calculated Cycle Time of 135 days, compared to the projection of 144 days.

"We had slightly lower Net Income Margin, 18.8% versus a projected 20.0%, but that deficit was covered by the higher velocity; our ROIA was 51%, versus the projected ROIA of 50%.

"With our baseline integrated building model, we had projected better business outcomes than the baseline outsourced building model, even with the projected disadvantage in pure operating performance.

"And, that turned out to be the case."

"Coincidentally, we had the same situation as Team 1, regarding resource performance. Resources A, B, and C all performed slightly below expectations; Process C was the system constraint. Resource D performed above expectations, so it, too, was often able to make up lost ground; again, Resource D can only work on what Resource C completed, and what was already in its own process.

"Whether we benefited from having inventory of six Projects-in-Process, I don't know. It would be interesting to track the performance of Resource C, to see whether it benefited from the extra inventory in front of it."

"How would you make that determination?", asked the intrepid, results-based consultant.

"I think one way would be to see how much of Resource C's capacity we failed to use", replied the superintendent assigned to Team 2. "Were there times when it could have performed more, if it had more to work on? In other words, did we starve the constraint?"

"Good point", she said. "Team 3, you had the Productive Outsourced Building Model. How did you do? What did you learn?"

"We exceeded our Revenue projections by 10%", said the CFO. "We were more productive; Productivity was 11.0, compared with a projection of 10.0. We were faster on the twins;

Inventory Turn was 3.1x and Cycle Time was 115 days, compared with a projected Inventory Turn of 2.8x and a projected Cycle Time of 126 days.

"We thought that performance would translate into proportionately higher financial outcomes, and we were not disappointed. Because there was a meaningful difference, we decided not to round the results to the nearest project; we stated performance as a range between rounded and unrounded results. Net Income Margin was in a range between 9.1% and 10.9%, which brackets our NI Margin projection of 10.0%.

"Likewise, our actual Return on Invested Assets was in the range of 28% to 34%, against our projection of 29%.

"In our scenario, Resources A, B, and D performed at or above expectations; those are all non-constraints. Resource C is our Constraint Capacity Resource, our CCR, because it has more work to perform than any of the other resources; in this game, Resource C performed slightly below expectations, averaging 6.9 tasks per round, instead of the expected 7.0 tasks per round.

"The fact that we completed 10% more projects than we projected is likely a reflection of the effectiveness of Resources A and B in making sure that Resource C always had projects on which to work; by their coordinated efforts, and because they had reserve capacity, Resources A and B rarely forced Resource C to be idle. Resource D had sufficient reserve capacity to always finish everything Resource C completed.

"This was the first instance in which price elasticity of supply and demand was imposed on operating statements; it made a difference, and that difference was reflected one hundred percent in the margins that were generated.

"It was also refreshing – as we have learned through many Pipeline games™ – to have a specific constraint to manage, instead of trying to manage everything. It is always a good reminder that balanced capacity is not the answer, that mandating even-flow through a rate of starts is not the answer.

"Last thought: we had decided to maintain seven Projects-in-Process as our WIP; at the time, we said we thought it was clearly necessary, we were not sure if it was sufficient. It seems like it was necessary and sufficient; now, the question is whether it was overly sufficient."

"We are about to find out", the intrepid, results-based consultant replied. "Team 4, you had the Productive Integrated Building Model. Tell us how you did. Tell us what you learned."

"First, I want to remind all of you – A-L-L-O-F-Y-O-U – about what my team said, prior to the game", said the sales representative. "We said that we thought 'necessary is more than sufficient', and that we think 'we can do more with less . . . less than what conventional wisdom would mandate . . . less than what a balanced capacity mental model would suggest'.

"Because, that is exactly what we were able to do.

"Like Team 3, we exceeded Revenue projections by 10%. We know that integrated building models will never generate the productivity rates of outsourced building models, because of the higher overhead reflected by our Operating Expense; nevertheless, Productivity was 5.5, compared to the projection of 5.0.

"Because we generated more Revenue with less inventory/work-in-process, we were faster-than-fast on the twins, the reciprocals that reflect velocity; our actual Inventory Turn was 3.7x and our actual Cycle Time was 98 days, compared with a projected Inventory Turn of 3.3x and a projected Cycle Time of 108 days.

"As a result, our business outcomes also exceeded projections. We also chose to express performance as a range between rounded and unrounded results. Net Income Margin was in a range between 21.8% and 22.7%, and that entire range of performance exceeded our NI Margin projection of 20.0%.

"Our Return on Invested Assets was in the range of 80% to 84%, compared with a projected ROIA of 66%.

"Our resource work rate/variation was different from the other three teams. Resources A and B performed slightly below expectations, and Resource C and D performed right at expectations. However, even with the lower-than-expected performance from the two resources that had to feed the constraint resource, we were able to do it, and we were able to do it with less work-in-process than what the P-OBM – the Productive Outsourced Building Model – carried. In fact, we did it with the same work-in-process as our cousin, the Baseline Integrated Building Model."

"What was the key?", asked the intrepid, results-based consultant.

"There were two keys", the sales representative replied. "The first key was making certain there were always two Projects-in-Process in the B3 task – that is, in the B Complete task – ready for Resource C to bring into the C process.

"In that way, Resource C could always obtain the maximum benefit from its highest possible work rate; and, if it didn't have the highest possible work rate in that round, it was in a better position to realize that benefit in the next round. And, if – for whatever reason – Resource A and Resource B were unable to always coordinate having two Projects-in-Process in B3, the time to miss would be when Resource C had work-in-process remaining in one of its tasks.

"The second key was to prevent any resource from multi-tasking. It was key for every resource to finish a specific Project-in-Process before it started working on the next – or any other – Project-in-Process."

"Very good", said the intrepid, results-based consultant. "Very, very good. There are any number of ways, any number of scenarios, from which to learn any number of lessons from the Pipeline game™.

"We have accomplished a great deal. The changes we made to the Pipeline game™ represent a huge improvement.

"Thanks to your input, we have completely revamped the operating statement, changed the way costs are accounted for and reflected, changed the way resources work. We have separated resource capacity from resource cost, by making them external, so that they reflect the true outsourced nature of homebuilding production.

"We have experimented with different building models, as a possible alternative to the accepted, near-universal industry practice of strip-mining the value stream. We have, again, looked at the benefits of improving the productivity of resources and improving the performance of processes.

"We have taken the game far beyond its origins as a simulation of manufacturing production. I wish I could say that I thought the necessary changes up all by myself, but that is not the case; these improvements would not have happened without your collective insight.

"The principles and disciplines that govern homebuilding production never change; don't ever forget it.

"Now, get out of here", she said. "Take everything you have learned – old and new – and put it to work for RB Builders, and for yourselves as the savvy, mutually-accountable, motivated team of stakeholders in the outcome, by producing better operating performance and higher economic return."

CPSIA information can be obtained
at www.ICGtesting.com
Printed in the USA
BVOW07*1957231016

465526BV00021B/32/P